THE DROVER'S
Daughter

Published by Brolga Publishing Pty Ltd
ABN 46 063 962 443
PO Box 12544
A'Beckett St
Melbourne,VIC, 8006
Australia

email: markzocchi@brolgapublishing.com.au

All efforts have been made to contact the copyright owners of images and any omissions will be corrected in future reprints.

National Library of Australia Cataloguing-in-Publication entry
Blackwell, Patricia.
The drover's daughter / Patricia Blackwell.
ISBN: 9781925367751 (paperback)
Blackwell, Patricia.
Drovers--Australia--Biography.
Women--Australia--Biography.

Printed in Australia
Cover design by Brolga Publishing
Typesetting by Alice Cannet

BE PUBLISHED

Publish through a successful publisher: national distribution, Dennis Jones & Associates & international distribution to the United Kingdom, North America. Sales Representation to South East Asia
Email: markzocchi@brolgapublishing.com.au

THE DROVER'S
Daughter

Patsy Kemp

Author's preface

When I moved to Melbourne in 1971, aged 20, I took with me a broad country Aussie accent that had people mocking me about the way I spoke. Dog was never "dog" it was "dawg" plus I had no dress sense and many other bushie quirks. I quickly learnt a port was where ships came in and a suitcase was for placing clothes for travel. As I spoke to others over the years they kept telling me I should write a book about my experience of being a drover's daughter. In early 1980 I started to put my memories on pieces of paper. Some of these I copied out and sent to Mum, Emmie and Mary to add to. Mum made an effort to fill in the gaps, Emmie did as much as she could but she found it very stressful to talk about our past life "in the long paddock" and Mary did not acknowledge my notes. With their help and the memories I gathered over the past 25 years I have managed to get a book together that I am proud to call my own.

I would like to thank my friends Leanne Jones, Tina Fry, Jacqui M and Deanne Berry for the many hours of "gramma" corrections, moral support and laughter as this book progressed over the past two years. Thanks girls.

A number of people both living and dead are mentioned in this book, which is my personal account. Any misconceptions that might be perceived are mine alone.

Patsy Kemp
patsykempdrover@bigpond.com
www.patsykempdrover.com

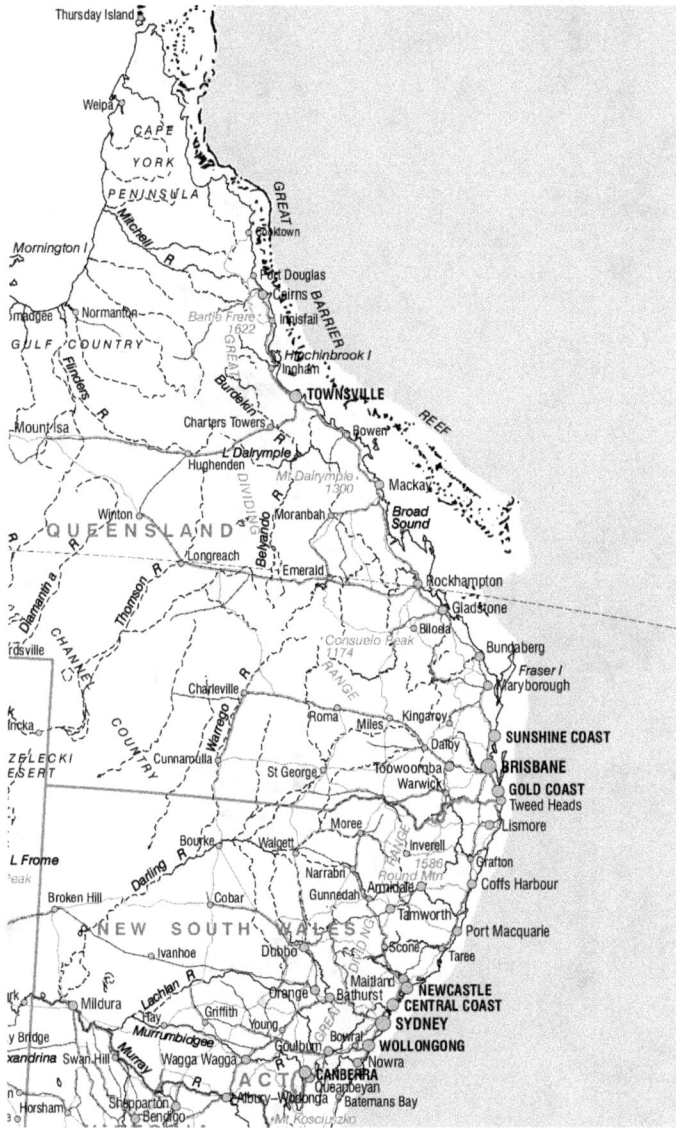

© Commonwealth of Australia (Geoscience Australia)

NARRABRI 1955

Traditional land of the Kamilaroi people

My earliest memory is one of terror. The flood. I was about four years old and it had been raining for days. Everyone was talking about an impending flood. We had been camping on the reserve beside the Namoi River but with the constant rain we were now short of food. The roads were mud and none of the vehicles could get into town. We were well and truly stuck. Dad decided that if the rain stopped he would catch the horse and ride into Narrabri to get some basic food supplies because if the river broke its banks we would be stranded for days.

Rain pelted down. There were nine adults and six kids and we huddled together trying to keep dry underneath a small lean-to Dad had built onto Mrs McCaw's humpy-like kitchen. Mrs McCaw, a friend of ours who also camped on the reserve, made us all a hot drink as the adults discussed our dilemma and the best way to tackle our predicament.

I was sitting on my eight year old sister's knee. Emily (Emmie) cuddled me as I sucked my thumb, mesmerised by the smoke from the adults' cigarettes lazily wafting out into the open to join with the rain. With the constant drone of their voices going up and down in heated discussion as to who had the best plan, I eventually drifted off to sleep. When I woke, it was early the next day and I was in the back of the truck along with my parents and siblings who had all slept there to keep safe. While they were still sleeping, brown muddy water rushed below. I could see it was halfway up the wheels of the truck and decided it would be fun to climb down and have a play while the others slept. I was having a lovely paddle in the flood waters until my mother woke up, horrified. She ordered my eldest brother Col, who was nine, to grab me before I was washed away with the debris in the swirling torrents. I couldn't understand why I got such a hiding when I was only trying to make mud pies for breakfast!

Thankfully, the attention was quickly diverted away from me and my wet muddy clothes as news arrived that the floodwaters had broken the bank of the river not far from where we were. Mum woke Dad. 'We have to move to higher ground,' she said in a panic.

Dad looked down at the depth of the water encircling the truck. 'Don't be so stupid,' he yelled. 'We're already on the only bit of high ground there is.'

We had parked on a sandy ridge that was used for a rubbish tip where the local night cart and odd job man, Teddy Small, had built a shed. Suddenly, a great wall of water appeared from seemingly nowhere and there was a desperate flight towards the shed. I was terrified. The biggest and eldest all had to lock arms and struggle against being swept away in the powerful current. I was perched on top of Col's shoulders and Dad carried Mike, aged three. My brother Les, aged five, and sister Mary, six, were being piggy backed by Alec and John McCaw. Emmie was struggling with the adults and at one point she lost her footing, slipped into the muddy water and began to float away. Dad quickly grabbed a handful of hair and tugged her to him. Mum shoved the blankets she was carrying under one arm and put her other hand around her eldest daughter's shoulders to help her along in the slippery, swift flowing water.

Sounds vibrated in my head: men yelling and swearing, the women and us kids screaming and crying, the rushing of the water, the absolute terror of it all. Teddy Small jumped onto the shed's roof and Col handed me up to him. He placed me a few feet away and then helped to drag the others, one by one, over the jagged, rusty guttering that was almost falling from the excess weight of rotted foliage and the heavy downpour now being thrust upon it.

I thought we were going to die. I sobbed uncontrollably as I watched drowned sheep, dogs, chooks and various other once-living creatures float past, tangled in the debris of the raging flood waters.

Beside the eight in our family there was Mrs McCaw and her five adult children, John, Alec, Colleen, Lucy, Mary; her siblings, Norman and Amelia; along with Teddy Small. Although we were grateful to be safely up above the water line, as the day went on, the rain stopped and the sun came out. The tin roof became unbearably hot and began to burn us. We were all hungry and thirsty as well. Dad told Col and Alec McCaw to go back to the truck and gather some food for us. They managed to fight their way there and collected as much food and water as they could carry. It only lasted one day and night as there were so many to share it with. Who would have thought a tin of green peas would be so satisfying?

For our safety at night, the adults put all the kids in the middle and they slept on the outside, even so Mike managed to wriggle close to the edge and if Alec McCaw had not seen him in time and grabbed him, he would've toppled over into the raging floodwaters.

John McCaw had two bottles of rum with him and he was in a drunken state the entire time, much to the adult's disgust. In the middle of the night, John rolled right off the roof and into the rapidly flowing water. He grabbed onto a log to keep himself afloat and began shouting. Dad heard him and immediately knew what had happened. He jumped in the water and grabbed him and, with the help of John's brother and Teddy Small, they managed to get him back onto the roof. It had been touch and go, as Dad was the only swimmer among them and it could have ended tragically for everyone, leaving Mum a 23 year old widow with six kids. Of course, the experience gave John a good fright and he sobered up pretty quickly. Dad angrily grabbed his bottle of rum and hurled it as far away as he could into the water.

The shed shook with the power of the water rushing by. The strength of the howling winds helped to make the shed feel more unstable and we were all afraid that it would collapse. My parents also feared losing the truck, which housed every possession we owned.

After a long eventful night, everyone breathed a sigh of relief upon seeing the police water boat approaching. They couldn't take us with them but at least they were able to organise some food for us later in the day. We were stranded on that roof for two days and nights before the water finally receded. Then the mammoth task of cleaning up began. The water had covered two feet over the back of the truck but fortunately, my parents had stored most things up as high as they could, so we didn't lose much at all.

The McCaws weren't so lucky and lost most of their possessions. Fed up with the camping-style living arrangements they had endured over the years, the girls decided to move into town but Mrs McCaw, her sister Amelia, and the men stayed where they were and started over again. It is a big move for bushies to settle down to living in a town, surrounded by people and houses. They become comfortable with their bushie lifestyle, free from noisy, nosy neighbours. They love to sit in the shade of a tree and smoke and yarn away the day when they are not working.

Most of Mrs McCaw's chooks and ducks survived the flood but sadly, three of their dogs drowned. Eventually, the McCaws set out their camping area like a village with tents and caravans in a circle around their underground bore that had a hand pump. Any leftover water made a nice handmade duck pond in the centre of their camp. This pond was very shallow and about 10 feet across. Ducks and various fowl life loved the smelly water and the dogs enjoyed a swim in it on a hot day.

One day Dad was teasing Mum and she retaliated by hitting him and then racing off out of the camp area. She ran around the tents and caravans, ducking and dodging the barking dogs that were straining to the end of their chains, wanting to have a run around too. Dad raced after her and caught her up into his arms. He carried her yelling and screaming back into the camp and held her over the pond. He pretended to draw her back to himself and then suddenly, tossed her fully clothed into the smelly pond. Ducks and chooks ran everywhere, feathers flying, dogs barked madly, and the adults watching this side show were all killing themselves laughing. We kids thought it a great joke. Mum squatted in the pond for a few seconds, a bit shocked. When she crawled out, dripping wet with mud, muck and green slime, she cursed Dad, which everyone thought was funnier still.

Stirring the pond up made the stink more prominent in the immediate area and Mum stank for days afterward. The rank smell was very hard to get rid of. Eventually, she started to smell like our Mum again – cigarette smoke! For years afterwards, it became a joke to make a wide circle as you walked past Mum to remind everyone of "the day Mick threw Beryl into the duck pond". Dad often threatened to throw her in the pond again but never did. I think he may have paid a high price for that bit of high jinks.

A few days after this incident, Mike was toddling along trying to catch a puppy. It swung away from the edge of the duck pond and Mike could not stop soon enough and fell in face first. John McCaw saw it happen and promptly pulled him out. Mike did not seem to smell as much as Mum had but being small, maybe it was because he didn't stir the sludge and muck up from the bottom of the pond.

The Namoi River ran past Wee Waa and we quite often camped on the banks when waiting for a job to come along or just resting the stock and the stockmen. There were plenty of ghost gum trees for shade and of course water and grass for the horses. It was lovely to lie in bed and just on dawn be woken by the cacophony of the kookaburras doing their early morning call. The galahs would start squawking and the crows arrrk arrrking. If we slept in too long, Dad would say they were laughing at us for being silly enough to stay in bed. On rare occasions we saw small flocks of black cockatoos and they squawked as they flew overhead rather than when they were resting in trees. The white cockatoos were more prevalent and a more common sight to us.

This area was abundant not only with bird life but with goannas and lizards. One of our games was to try and catch a goanna, so as soon as we saw one, we would give a shout and then the chase was on. Of course, we never had a chance of catching them. Goannas' bodies are quite close to the ground but when they are frightened or being chased they stand high on their legs and run very fast up the first tree they come across.

Dad enthralled us with a story of a man who was out fencing and he disturbed a goanna. His dogs gave chase and the goanna ran up the horse's leg, mistaking it for a tree. The horse was in the shafts of a cart that still had several fence posts resting in it. The horse got such a shock, he bucked and galloped off, the cart jumping all over the place, fence posts bouncing out of the cart and with the goanna standing full height on the horse's mane, as if he was steering the horse. The dogs gave chase and the horse eventually arrived back at his owner's house a few days later. The cart was missing but he still had the harness on with some bits missing and a few scrapes and scratches on him.

Dad was hired by Dick Holsbourne, a station owner in Wee Waa, to muster stock, do some fencing, horse breaking or whatever needed doing. Dick had lost an arm in the war but he generally managed quite well. After they mustered the stock, the sheep were shorn and Dad got the job of taking a large mob of sheep out on the "long paddock". The trip lasted quite a few months and we worked for Dick on and off for a few years after that. Dad became Dick's drover of choice, which was a compliment to Dad.

Often when we were droving along steady with the sheep, Dick would turn up with some newspapers, bread, our mail, fruit and lollies for us kids. He would always spend a bit of time at the camp talking to Mum and with a cheery, 'Hooray,' he would get back into his car and continue on to the men travelling with the stock. He was a wonderful gentleman and we loved to see him, not only for the lollies!

When the stock came into the camp at the end of the day they had to be settled into a sheep break for the night. A sheep break is constructed by rolling out "ring lock" wire into a half moon shape onto a permanent fence line. Posts are then put into that shape and the wire tightens out as you slowly and evenly use more star pickets to hold this temporary fence in place. You leave about 20 feet or more open at one end where the sheep can come in and then that end is closed up for the sheep to bed down for the night. If you were lucky and had a corner to attach the sheep break to, the job was halved.

Mum and us kids were quite often used as spare dogs to help guide the sheep into the break if the sheep were a bit skittish or if they spread out too far. They had a habit of going around the sheep break rather than in it and there was always a rogue sheep who tried to rush past and would jump over us. I was often knocked over by a galloping sheep but rarely got seriously hurt. It was no use crying to Mum, as she would comfort me with, 'Go on, have a good cry. The more you cry the less you pee.'

My older sister Emmie was far more motherly or at least showed more sympathy than Mum offered. Sometimes I'd cry behind a tree with self-pity, thinking I was so hard done by.

One day a storm was brewing so it was all hands on deck to get the sheep into the break for the night before the storms hit. Mum and us kids were helping when a sheep broke away from the mob and made a dash for freedom. Dad yelled at Mum to go after it. About an hour later Mum staggered back into the camp, red faced, sweating, panting and obviously furious, with no sheep in sight. Dad made the mistake of asking where the sheep was.

'I caught the f...ing mongrel and hit it on the head. Next time you run after the f...ing thing yourself,' she yelled. Mum had chased the sheep for ages and it eventually tangled itself in the fence wire in a panic to get away. Exhausted and seething with anger, Mum then had a long, weary trip back to the camp. By the time she arrived, the rest of the sheep had been penned but the men had not lit a fire or done anything towards starting dinner and this made Mum even angrier. There were six kids and three men still to cook for as well as finishing setting up the camp.

The movie *The Ten Commandants* came to Narrabri and Mum

decided she would take us all to see it to, 'Get some religion into us.' Before the movie, we were all lingering outside, possibly for the parents to have a cigarette. Smoking was allowed in cinemas in those days. I thought I was standing by Dad's legs and felt quite safe looking around at the night lights and things going on in the street. He walked off and I followed before realising none of my siblings were with us. I looked up – this was not my Dad! Horrors, I had been kidnapped! I got such a shock seeing this strange man who was unaware of me trotting beside him. I looked back and saw my family half a block away. Luckily, I had not even been missed, as it would have meant a good hard smack on the bottom for wandering off. *The Ten Commandants* was a long, exciting movie and was so good I did not fall asleep at all.

Around the same time, *The Big Chief, Little Wolf Circus and Buck Jump Riding Show* was in town. Dad was half-tanked and he walked up to the Indian sitting in front of the big tent, wearing his feathers and other Indian clothing.

'You're not Big Chief Little Wolf,' Dad said.

The Indian replied, 'If I'm not Big Chief Little Wolf, I'm having a hell of a good time with his old woman!'

Dad fancied himself as a bit of a rodeo rider and for some reason chose to ride a donkey. His mates hoisted him up onto it and the donkey ambled off slowly. In Dad's drunken stupor, he fell off landing face first in the hay that was on the floor. One of his mates bent over with laughter pulled him upright again. All the people in there were laughing and Dad proudly straightened up, grabbed his hat off his mate and waved it around to the crowd. He staggered outside with a silly grin on his face ready to face the full wrath of Mum. On the way home, Dad was driving all over the road.

'Keep on the bloody right side,' Mum said.

'Whash shide of the road?' Dad slurred.

'Well keep to the bloody centre then,' she snarled.

'Where ish the schentre of the road den?' he replied. It was even scarier when he got to the bridge. From the back of the truck where we were, we could see down into the river bed, water glistening in the moonlight. I was petrified. From the front of the truck we could hear Dad singing a ditty and Mum going ballistic at him. Nothing worse than a drunk when you are sober I guess. We arrived home safely and were delighted when Dad tried to climb into the back of the truck and fell over backwards. He was laughing so much he could not get up so Mum threw a blanket over him and left him there for the rest of the night! She was fed up and gruffly told us to get to bed.

SURAT 1955

Home of the Mandandanji Aboriginal people

We were going into a sheep station called *Moolah* a few miles west of Surat and fifty miles out of St. George south east Queensland to pick up 3,000 head of Border Leicester sheep to drive them to Nyngan, in north western New South Wales. We found the station entrance, a sagging post that had a rusty tin bucket hanging off it being used for a mail box with the name of the station written across it. The front gates were what the bushies called COD. This meant you could either "carry or drag" them opened and closed. The long lonely dirt track into the station homestead followed the fence line for miles accompanied by a wire for the telephone. The line went from tree to tree and in many places, hung so low, it nearly touched the ground. The whole station was very run down and had a look of desertion about it. To add to the gloominess, the ground was bare of any grass or green herbage of any sort and the small gum trees and bushes looked sad, dry and miserable. Dad took the stock, counting them out of the yards, agreeing with the owner on the amount we now had charge of and signed off on the job.

We eventually arrived at Mungindi, the New South Wales border crossing in that area. The river was high and water was over the stock crossing, which made it impassable for stock, so we had to go over the actual town bridge. This meant taking the stock between the houses on each side of the road to reach the bridge. The sheep did not mind except when a skinny mongrel dog ran from under a house and barked at them. Some sheep got into the house yards that didn't have fences, nibbling on the lawn or getting into the flower gardens for a quick bite before our dogs chased them out with a whistle from one of the men.

None of us were ever too young to help in any way that we could. Over time, Mum had collected tin lids to make rattles. A hole was made in the middle of the lid and about ten lids were threaded onto a six inch piece of stiff wire that was twisted into a circle. We rattled these to keep the stock moving at a brisk pace. With tins rattling, dogs barking, and people shouting, the town's people came along to help as it was a novelty to see a large mob of sheep passing through.

The stock was herded to the beginning of the bridge but balked at going onto it. Dad grabbed a sheep and carried it across and it promptly turned around and ran back into the mob. He grabbed another one, tied a dog chain around its neck and carried it part way over the span in view of the other sheep, tying it to the railings. He then came back and shooed some more along. It was hot dusty work, as we were all kept busy moving the 3,000 sheep onto the bridge that didn't have guide rails on the approach. My brother Col rode his horse across to stop the traffic on the other side where cars had started to bank up. Col had to stand his horse in front of the first car to stop it from edging into the sheep as they came over. As the last of the mob bunched up onto the bridge to cross the Barwon River into New South Wales, Dad lifted me up and put me onto a sheep.

'Hang on love, ride the bastard over the bridge,' he said.

Being fully instructed, this was exactly what I did. The ride was rather bumpy as the sheep did not like me on its back. It tried to run between its mates and jump on the ones in front of it but with my little hands clinging on to the wool and my legs firmly dug deep into the sheep's woolly sides I managed to stay put. I was probably the first person, maybe even the only person, to ride a sheep over the border on the Mungindi Bridge! I could hear my mother screaming at Dad to get me off, but I think he was laughing too much to hear.

When we came off the bridge on the opposite side, the sheep I was on started "baa baa-ing" and ran crazily amongst the mob. After a few seconds of this, I let go of the wool and fell off backwards. I picked myself up, dusted off and stuck my thumb in my mouth for comfort. Emmie rushed through the mingling sheep and gave me a cuddle and over her shoulder I could see Dad striding back to the truck. I poked my tongue out at him, feeling bold and extra safe, as he had his back to me. Dad had often put us on a sheep when placing them in the sheep break at night and we always fell off quickly.

The Border Leicesters were short and wide and as this mob was near full woolled, it made it easier to hang on. My siblings admired me for my ride and were a bit envious. For years afterwards, my bravery was spoken of to any friends and acquaintances who would listen. I think Dad had expected a frightened sobbing child waiting for him over the bridge. It was very high and it would have been a long drop down to the water for a four-year-old if I had fallen off that sheep.

Most roads we travelled on were rough and dusty, either bare dirt or gravel. Things in the back of the truck quite often broke. We did not have crockery until we got a caravan in the late fifties. Up until this time, we ate off tin dishes and later on enamel plates and drank from enamel mugs. We had a lot of plastic dishes too that lasted longer then the enamel ones. Enamel chipped badly but we still used it. If we were short on mugs, the stockmen would use the cups off their quart pots to have a drink. The cooking was done in the camp ovens or aluminium saucepans.

It was not always convenient to bake in the camp ovens. A good fire had to be made with coals under the oven and on top of it and that was not possible if there was not a lot of good quality wood around. If we knew that there was a shortage of wood at any upcoming camps, we would stack some in the back of the truck and use it sparingly. It does not take a lot of wood to get a billy to boil or cook up a pot of boiled potatoes and pumpkin or heat a tin or two of peas. A small fire would be lit and this was normally called a "black fella fire" or a piddly attempt at making one.

To light a campfire in high grass we had to dig a hole about eighteen inches deep, depending on how hard the ground was. The soil from the ground was stacked on the opposite side from where the wind was blowing, so it offered more protection if the wind was very strong. The smallest flame could turn into a raging bushfire and we were all aware of this. The flames would literally roll along the ground hungry for substance.

It was common to see at least one of us squatting near the fire, one hand holding a forked stick with bread on it over the hot coals and the other hand held in front of our face to keep the heat off. If the fire was big, the stick had to be a couple of feet long or you cooked your face while browning the toast.

Dad had made a tucker box but it had disintegrated over the years, so he bought a green tin travelling trunk. Though it did not keep the ants out, it certainly kept out flies and the other pests that we preferred not to share our food with. In this box we carried all the things we needed on the table: hot sauce, tomato sauce, dry milk in half pound tins, mugs, cutlery, plates, syrup, Vegemite and opened tins of jam.

When this trunk was required, it was lifted down onto the ground and then carried to the closest tree that had half decent shade. It was actually quite heavy and as the truck was rather high, it was never my job, though I was quite capable of jumping into the back of the truck and pulling it to the door so Emmie and Mary could carry it. If the shade of the tree was sparse, it was shared by all – dogs and humans. We had tables on and off over the years but if they broke while we were on the road, miles from anywhere, then we made do without until a new one was bought. When we did not have a table, we ate picnic style and we learnt not to drop our bread.

We lived with the constant knowledge that water was not to be wasted. We had two canvas water bags on the side of the truck and hung an enamel mug off a hook to use when we needed a drink. But if no one was around to see us, it was easier to just tip the water bag and drink straight out of it. We all hated to see others do it but we appeared to have no qualms about doing it ourselves. We were only allowed to add three mugs of water to the small cream enamel dish we used for a hand washing basin. This same water was used by all of us until it was quite dirty, then it was thrown out and more water added. We always left a hub cap full of water for any dogs coming into the camp, otherwise they made straight for the hand washing dish. A cake of soap was with the dish and quite often it was left in the bowl to become a big blob of jelly (which was always someone else's fault!) and an old threadbare towel was hung nearby.

Petrol was very dirty and Dad kept an old pair of Mum's panties that he used to put over the nozzle of the petrol hose to filter the fuel as it went from the bowser into the tank. He used to take great delight in embarrassing Mum by putting his head in the window and saying, 'Take your panties off love and give them to me so I can filter the petrol.' He did not care who was walking past the truck as he said it and he never got sick of the joke at Mum's expense. She would always take the bait and be mad at him and then sulk for the rest of the day.

Dad and Mum went to Collarenebri to do shopping. Dad went to the Stock and Station agent to collect any messages for us, as our mail was always directed to the next town, and then continued to the pub. Mum did the grocery shopping and posted any mail. After finishing the shopping, she sat in the truck cabin with Mike. We "young ones" were quietly sitting in the back of the truck. Mum eventually got fed up and walked to the door of the pub. She poked her head around the door and spied Dad, then said to one of the patrons, 'See that red-headed bastard over there, well you go and tell him I want to see him.' Luckily Dad was a happy drunk and took no offense and back to the camp he drove.

Another time we were camped close to town, Dad was once again in the pub and Mum had Mike with her and no money. She decided to 'leave the red-headed bastard in the pub' and walked out of town, carrying Mike who was eighteen months at the time, back to camp. When she arrived, the blue cattle dog would not let her in, so she sat on a nearby log and cried and cried. She was tired, hungry and very upset as she knew she would possibly have a long wait until Dad got home. In her annoyance, she had left Emmie, Mary, Les and me in the back of the truck and she didn't know what we were up to. She was too tired to walk the couple of miles back into town carrying Mike. When Dad eventually arrived back to camp half-tanked, he asked her why she walked and she told him in the only way he appeared to understand, with much screaming and bad language.

Mum never bothered to walk back to the camp again. A couple of times she risked driving the truck back herself but she hated driving in town. Instead, she would annoy Dad so much at the pub, he would eventually give in and drive us all back to the camp. In those days it was frowned upon for women to be in pubs, hence her calling from the door. Sometimes the other patrons would tease Dad about Mum and the kids wanting to go home. He always made her wait an hour or so but then he would leave, often singing her a dirty ditty to get her into a good mood again.

We were camped on a common where there was plenty of green grass for the stock so Dad called a holiday. Some of the leather bridles, halters and hobbles needed fixing so he decided to do that. I used to love watching him get the hemp, rub it in with wax and roll it. This made the sewing waterproof and extra strong. When possible he would just push his awl through by hand, but if the leather was extra thick or several thicknesses, he would use a hammer and gently tap the awl until it was through all layers. He would rarely get anything fixed professionally, preferring to do it all himself. He also had a leather punch that made various sized holes in the leather. We often borrowed these tools to play with, making holes in anything we could and we were very careful to put them back into the bag in the crate where we got them from. Dad taught us to always put things back where we got them from, so next time they were needed we could find them. He preached this to us all the time.

One day we met up with Tom Bunyan's delving team camped on the side of the bore drain. With stock walking in and over the drains, the drains eventually filled in with dirt and debris, stopping the water from flowing into the next paddock or onto the neighbouring station so their stock would not get any water. When the drain was clogged entirely, the water was wasted and would flood the surface of the land. This was okay in one respect as the grass grew in abundance, but it was frowned upon if it was allowed to go on. To keep the bore drains open, a bore drain delver was hired to clear them. In this team was Tom Bunyan, his wife and daughter and her young child and another hired man. They bore drain delved in the St. George area. It was wonderful to see the huge draft horses walking along pulling the delving machine, two to four each side of the drain pushing all the mud, water and silt up out to the sides and over the edge. In the mud would be various sized "yabbies" as we called them. We would collect them from the mud and water, throw them into a bucket of boiling water and eat them with gusto. What a treat for us all! They were very rich though and we younger ones quite often got sick from eating them.

The delver's camp was set up similar to the drovers. They had a fair-sized tent put up as a kitchen and living area and they also had their own personal tents away from the main tent. Their living conditions were much the same as ours, although they did not move every day, which had to be a good thing. I was amazed to see they had chooks scratching around, clucking as they looked for seeds and bugs. When they were called to camp, they came running and jumped into their crates. When the Bunyan's moved camp the chooks were carried in these crates that were packed under the wagon. It was quite usual upon arriving at the next camp to find an egg or two in the crate.

The Bunyan's also did earth tank making, using a wooden blade with the draught horses pulling it and ploughing fire breaks around station boundaries to stop any grass fires from entering the property.

If you were a willing worker there was always a job going.

On this trip, we met another droving family called Wilson. There were quite a few Wilsons who were drovers in and around St. George. Over the years we met most of them and if they were driving past they would always call in to catch up with the news. George and Bertha with their children Georgie, Tommy, Julie (Maud), Shirley, Tony and Betty. The family was rather unusual because although the kids were around our age they spoke like adults. We were in awe of them as they all used bad language and swore like troopers, which we were not allowed to do. They also smoked. The Wilsons senior wanted us to call them by their first names but our parents insisted on us calling them Mr and Mrs or Uncle and Aunty. Aunty Bertha was a warm loving adult and we Kemp kids adored her. She was always ready with a cuddle and a kind word. One day we were camped with them in St. George and she was hand washing for the family and got a bit fed up with the wash load.

'That's it,' she said, 'they may not be clean but they will smell fresh.' She dipped them all in water and put them on a nearby fence to dry in the sun, then sat back and had a nice cuppa.

Around this time in the mid-1950s, a dress came into fashion called a Muu Muu. The Muu Muu was a shapeless dress that hung from the shoulders and the men hated them. They had a split up the side to about the knee and they were very roomy. Pregnant woman could use them to the final day and no bump would be seen. This fashion did not stay around for long. One day Crow (Roy) Wilson grabbed the bottom of his wife Marion's Muu Muu and ripped one side of it right up the seam to her armpit. He did this in front of everyone. We kids were horrified, but the Wilson kids thought it a huge joke and so did Dad and Crow. Marion told Crow off, but he knew she would not buy any more.

My older brother Col had a camp stretcher that he slept on under the stars. Each morning he had to roll his bed up and leave it on top of the stretcher near the back of the truck. Apart from a thin mattress and his blankets, in the winter he had a tarp that was thrown over the bed entirely that kept him dry and of course warm. Our blankets were ex-army and very thin. If we had any men working for us, this is how they slept also, outside on a stretcher. Col would place his hat over the top of his boots to stop creepy crawlies getting into them. He once had an incident where his foot would not fit entirely into his boot. He stood up and forced it in in frustration but when he turned his boot upside down and shook it, out fell a dead squashed green frog!

It was a bitter cold winter with severe frosts and Col could not keep warm at night. In the morning his bed would be white with ice. Mum pulled two potato bags apart and joined them together and then hand sewed an old double blanket around the bags. Col was quite pleased with his new "wagga" blanket and it kept him nice and toasty. Most country women didn't have a lot of spare money and had to use their ingenuity to keep their family fed and warm.

Christmas 1955 was spent camped on the Namoi river bank. I went shopping with Mum and she bought all these lovely big trucks, dolls and other things that would delight any child. I was wondering which gift was going to be given to whom but was not allowed to ask. Mum used to say, 'You are like the bird on the biscuit tin, seen and not heard.' This was referring to the Arnotts biscuit tin that had a colourful parrot on the side nibbling a biscuit that it held in its claw.

Dad bought a wooden crate of Orbell's soft drinks that was put in the river to keep cool. An extra-large watermelon was placed in a bag and tied to a tree root and also placed in the water. Christmas morning came and I waited anxiously for all the delightful goods that Mum had bought for us. What a dreadful disappointment to discover the presents were given to the Holsbourne children or posted off to Aunty Anne's children. I cried and cried over the doll that I had thought I was going to receive and love like no one loved me. Instead we received the same gifts we received each year: a colouring-in book and pencils, reading book, a small tin of toffees and some much needed clothing including a new swim suit.

Swim suits came in handy, even though none of us could swim. We could jump in bore drains and dams for some fun, always staying in the shallow end under Mum or Emmie's watchful eye. On occasions the stock trough was a handy bath. All the stock troughs had a windmill near them so they could pump water into a tank and then into the trough with a bore cock in it to stop it from running over onto the ground. If the tank overflowed it ran into a dirt dam as a secondary water supply. Quite often the trough would be full of dirt and slime and really green. We had to first clean out all the muck then refill it before we could get into the water. If a trough did not have a bung to unscrew at the end, or we could not open the bung because it was too tight, we would use a broom to get the muck off the sides and bucket the water out by hand. A nice clean trough was for our pleasure but the animals benefited too, with fresh clean water to drink. This was not done on a regular basis as some places didn't have the water to waste.

Sometimes we would dare each other to get in the trough before the stock arrived. We would sit quietly in the water while the stock had a drink close by. We would try and sneak in a pat but we never managed to connect with a beast as they were not that tame. The horses would look at us curiously, as if to say, 'What are you doing in my drinking water?'

Wash day was a big event that had to wait until we were camped near any sort of clean water. A bore drain, dam, river or ring tank would do. If we were camped near a river, no matter how steep the bank was, we had to bring water up to the camp. I can remember one particular time, we were camped on the banks of the Namoi River between Narrabri and Wee Waa. The bank was quite steep and we had to carry the water up from the river in tin buckets. The buckets were used-kerosene tins with wire handles. Our hand knitted jumpers and cardigans had to be hand washed as well, these had to be drip dried and laid flat on nearby bushes – or on Col's bed.

DIRRANBANDI 1956

Traditional land of the Kooma people

Dad was offered a job at Dirranbandi to manage a station for a month while the manager, Jack Smythe, went on a well-deserved holiday. The property was owned by the Australian Pastoral Company (APC), one of the biggest landowners in Queensland at that time. Dad had a lot of work through the APC over the next few years.

This station was full of lignum, a plant native to Australia. The homestead was totally surrounded by it. Mr Smythe gave Dad a mud map of the station and a quick look around including which cows to milk and a list of jobs to do. Dad and Col were out every day doing the work while we had the pleasure of living a life of luxury. We even had a toilet, rather than having to go behind a tree or bush. Unfortunately, the toilet was the old "long drop", meaning the hole was about 6 feet deep and the toilet seat was a thick plank of wood with a hole cut in it. It was one size fits all and was not meant for small bums. When we little ones sat, we would end up with our bums dangling down into the hole. It was huge to a five-year-old and I hated it with a vengeance and eventually refused to go in there. I had nightmares over this toilet for years and I wet the bed many a time dreaming I was falling into that great gaping hole. To compound the problem, the toilet was full of spiders, they were also in the great gaping hole under the toilet seat.

The station had a public road going through it and instead of having gates or grids at the fences, a dog was tied to each side of the road. They would run out barking at the stock to stop any that tried to come through. While there, two of the dogs were mauled to death by wild pigs that obviously wanted to go through the gate and did not appreciate the dogs' efforts trying to stop them. This also gave me nightmares for years as I heard the full graphic story when Col came back and described how the dogs had been eaten by the pigs. I was a sooky sensitive child, and the smallest thing would set me off crying and sucking my thumb.

Dad found milking the cows difficult as some did not take kindly to being milked by a stranger. He had to practically tie them up when he milked them and when Mr Smythe and his family came back, they found that Dad had inadvertently broken in several new milking cows. Dad had milked all the cows that had hung around the shed as he had forgotten which cows were supposed to be milked and which weren't. The non-broken in cows would have put up a real fuss at the indignity of being milked. No doubt kicking out their back feet and trying to butt Dad away.

While in Dirranbandi we did many trips in other towns. At some stage Dad met up with a couple in Dirranbandi he got on really well with. Tom and Merle Crumblin had several kids and we were all around the same age. Tom was the rubbish man, wood carter and the night cart man. He had a tip-up truck and went from house to house picking up their rubbish and dumping it into the back of the open truck. We loved to pick over the rubbish if we happened to be at their place when he came home for a meal but he was not so popular on the night cart days. Phew the pong!

The night cart job was to go door to door. At the back of the toilet was a door that was big enough to pull the full can out and push the new empty one back inside. Tom would have a potato bag slung over his shoulder and he would hoist the full can onto his covered shoulder and carry it back to his truck. If the can did not have a lid on it or was a bit full, it would slop over onto Tom's head and shoulders, though Tom always wore a hat. This smell would permeate everything and when home on these days, Tom was only allowed on the veranda, not inside the house. As a side note, if the cans were full before the night cart man was due to come back, then the house holder would dig a hole in his back yard to bury the muck and this made good ground for growing vegetables later on. Tom said often someone would be sitting on the toilet and he would greet them with a, 'Good mornin darlin,' if a woman or 'How ya going mate?' if a man. If he did not know the sex of the occupant, he would say a cheery 'Good mornin.' In a town the size of Dirranbandi, everyone knew each other and quite possibly, someone knew their bowel habits too.

There was no toilet paper in those days for the "poor folk" so they would cut up newspapers and magazines into four inch squares and place a corner of it onto a twirl of wire that hung on the back of the toilet door within easy reach or on the side wall of the "dunny" as it was called in the bush. A luxury we never had was to gather the soft tissue paper wrapped around apples and pears and use that. Quite often a vine of some sort would grow over these small buildings and it would become a haven for snakes, frogs and any living creature that fancied a warm or cool spot, depending on the weather. Most of the time the dunny door was kept closed, if not any animal could go inside. Often as you entered you would be greeted by some creature wishing to get out! If it was in the dark of night it was very frightening for kids and women. There was rarely any privacy going to and from the outside dunnies and neighbours would have a natter with you while you were trying to have a quiet pee.

Tom had a large council paddock on the Noondoo Road where he buried the sewage waste. The day before collection he would dig large holes in this paddock, much like a corner post size and the sewage cans were tipped into these, then filled in with dirt. One day Tom decided to bring his boys and the Kemp boys to help. With the truck filled with cans, the boys sat on top of it and yelled, cooeed and yahooed all the way as Tom drove out to the paddock. The first thing Tom said to the younger boys was, 'Don't run around, you will fall over and get hurt.' Tom carried a can and tipped it into a hole. Col and John carried cans between them. Les never listened to anyone and ran around the paddock, jumping over filled and empty holes in the ground. He tripped over a pile of dirt and splat, went head first into a newly filled sewage pit. He let out a scream, stood up and shook himself like a dog, with brown slime and pieces of paper dripping off him. Tom glared at him. 'I bloody well told you not to run around. Now see what you've done.' He went to the truck and retrieved a potato bag, wrapped it around Les and sat him down near a truck wheel, refusing to pander to him by not taking him home immediately. Mum was not amused when he arrived home in that state.

At that time Lonnie Donegon's song, *My Old Man's a Dustman*, was popular on the radio and we all knew the song off by heart and when driving into town to visit with the Crumblins we lustily sang it. I am sure we could be heard leaving the campsite about two miles out of town. We had a transistor radio at that time that we called a "wireless" and it could only be used on very rare occasions. Dad listened to the news occasionally, any boxing matches that were being broadcast and the Melbourne Cup. If he was not around and we could get reception, Mum would sometimes let us listen in.

When it was rubbish collection day and Tom came home for dinner before dropping the full load off at the tip, we would jump up on the top and have a good rummage. This scared Mum silly as she was frightened we would fall off and hurt ourselves. The younger ones were banned from getting on the truck, but I am happy to say that no one ever fell off in our foraging.

While John, Arthur and Les were rummaging around one day, John found a big doll fully dressed. She had a lovely printed pink frock, a waist band, real hair and a pair of little shoes. John dusted it off and threw it down to me. What a treat! He had a sister not much younger than I was, but of course she was at school. Mum was never too impressed with our rummaging as she was afraid of diseases but we loved it and lots of good stuff was found.

I loved John Crumblin and wanted to marry him when I was older. I always wondered why the doll had been thrown away like that, as it seemed brand new. Sad to say the doll went like the rest of our toys, lost, worn out or just plain forgotten and left behind at a campsite. Often one of the older brothers would tease us younger ones by holding our prized possessions out of reach for us to try and grab or they would pull the toy apart for a "tease", so we never had anything precious for long. We two youngest ones were small and the older boys, when we tried to attack them in revenge, would just put their hands on our heads and hold us at arm's length so we could not reach them. I would get so wild I would cry, then go to a quiet corner and suck my thumb.

Tom and Merle's house was on the Balonne River bank and they offered us the use of their yard to camp in, which Dad accepted a couple of times. At this particular time, the river was in flood and John, Arthur and Col insisted on swimming in it. Merle got into a ripe old temper. She screamed out, 'If you get yourself drowned, don't come crying to me.'

We thought this was hilarious and crowed about it for years. I remember her turning to Mum one day when she was very annoyed with her brood and saying, 'You got one boy, you got a boy; you got two boys, you got half a boy; you got three boys, you got no boys.' That puzzled me for years but eventually I understood the truth of it.

One day Tom upset Merle dreadfully. In front of us all, she stood on the front veranda and threw 20 pounds of potatoes at him – one at a time. She did not hit him once. We all thought it very, very funny, which enraged her even more. Another time we were there for dinner and Merle made a pot of tea. Tom took a mouthful of it and spat it out as it scalded his mouth. He complained, 'What did you do Merle, boil this water?'

We kids were taught to tell the truth but we learnt that telling the truth to an adult was not always a good thing. Merle made a cake and it was a bit doughy, but as we rarely had cake, we all ate with great gusto. When he'd finished eating, little Mike said, 'Thanks for the

cake, Mrs Crumblin. The tea was nice but the cake was a bit soggy.'
That got him a slap from Dad for bad manners.

Another time Merle gave her younger son Tommy, a hiding with
bamboo and he got a big splinter in his hand as a result. Dad tried to
take it out but it would not move. So Dad volunteered to take him
to the hospital to get it removed and before they left Merle said to
Tommy, 'If you tell them I flogged you with bamboo, I'll kill you
when you get home.' At the hospital, the first thing Tommy said was,
'She flogged me with bamboo.'

Tommy always wanted to please and Dad was a real scrounge
who took anything offered to him that might come in useful. One
day Tommy found a washer that he thought Dad would like and
offered it to Dad, saying, 'Mr Tent, do you want a squasher?' Washers
became "squashers" for many years after that in our family and Kemp
quickly became "Tent". Being little blighters we would call Dad Mr
Tent to get a reaction, which was mostly a light kick up the bum or
clip over the head.

John Crumblin whipped up a homemade canoe and we all got in
it. He placed a drum each side of the canoe and held it together with
a rail about two inches thick to balance it. Getting into the canoe was
an art in itself. John and Col were at each end and the rest of us very
gingerly stepped in. As it was a tight fit, the older ones sat with their
bums on the edge and their feet in the canoe and we smaller ones sat
in the middle. Mum saw us out in the full flowing river, the canoe
wobbling madly, and she called us in. So with great disappointment
and trepidation we paddled back to the bank. We all received a good
hiding for doing that as half the Kemp kids still could not swim at that
stage. Merle locked her kids in the chook pen to punish them. We all
considered Mum a real "spoil sport" and that we were allowed no fun.

In the Crumblin's large yard they had an old empty tank on a high
stand and we enjoyed climbing into it to play. It would shake and
rattle and it's a miracle that it didn't ever topple over and kill us. The
yard backed out onto open spaces and kangaroos often came in and
when they did the boys would run like mad to try and catch them.
One day Les caught one by the tail and he flew one way and the roo

went the other way… who got the biggest fright is debatable.

We often camped on the reserve some two miles out of town. Mum much preferred living there as it gave us more freedom and Dad and Mum more privacy. Mum was a screamer and she had the liberty to shout at us kids to her heart's content.

The Culgoa Common was a camping spot for drovers and swaggies. It was a good place to camp as there was shade, water and grass for the horses and we did not have to worry about the dogs barking and being a nuisance to anyone in the township. Nearby was a farming family called Dean, we could shout across the river to them and they would come over and play with us. When the river was low in water and there was a tree that had fallen over we could walk across for a visit. As Mum kept a tight rein on us, the Deans had to do most of the visiting to "our" side of the river. They were very friendly and any drovers who camped in this common got a visit from the Deans. The two Dean children still living at home, Leithy and Leroy used to ride their bikes to school in Dirranbandi. When we had stock, our camp was right near their gate and it was great to meet these kids and have a chat before they biked to their house up the road a bit. They were in the older kids' age group, but we all enjoyed their company. We looked forward to camps like this as we met so few kids our age group. The family had two Alsatian dogs Pedro and Kim and a horse called Tony that Leithy rode. When the Dean family heard we were fairly close on a trip with stock they would travel out on the stock route for a visit and the adults would play cards into the night and we kids would play games in the dark.

Their Dairy was called *Riverside* and it was a couple of miles out of Dirranbandi on the Bollon side of town. Mrs Dean carried the milk into Dirranbandi every day with a horse and cart and after a while they bought a secondhand ambulance that their daughter Mackie drove for them. Neither Mr nor Mrs Dean could drive.

Mackie was thirteen and worked for the Pippos who owned the local *Café Deluxe* in town. Our parents would often go into the café and Mackie would serve them their mixed grill. We would be out in the back of the truck and after they finished their meal they would buy a large serving of hot chips, which was wrapped in used newspaper, for our meal and the drive back to the camp. Col would get his share. He never seemed to mind being left behind to mind the camp but then he didn't really have a choice. At this stage he would have only been about ten years old.

In the earlier years, Mackie, Leithy and Leroy would ride their horses to and from school. The local Aboriginal children who lived near the Deans on the opposite side of the river had to walk to and from school, having no other mode of transport. After school was over there was a wild rush to get to the horses. Whoever got to them first, got to ride them home. When the Aboriginal children rode them, they would gallop up near the bridge, tie the horses to a tree and walk the rest of the way home. The Deans would walk to where their horses were tied and ride the rest of the way home.

Many years after this incident I met up with Leroy and he told me fondly, 'The little black bastards would pull us off our horses and gallop home to their humpies, if we were lucky, they would not bash us up.'

No grudges were ever held, he who ran the fastest got the horse. Eventually, the teachers got cunning and would let those riding home leave school early, cutting out the wild dash to the horse paddock.

Being on the *Culgoa Reserve* we could paddle in the water and Col, Les and Emmie learnt to swim there. The Culgoa River was a muddy water hole most of the time and was full of leeches. One day Dad and Mum went shopping in Dirranbandi with all the kids except Col, who was always left at the campsite to mind it, in case someone came along to steal things. We called in at the Crumblins for a visit and Merle asked us all to stay for dinner. Mum said no we could not stay as Col was waiting for dinner out at the camp. So Merle told John to race out and get him and John did that and they both then jogged the two miles back into town. It proves how healthy they were. John had one of his big toes missing, he had accidentally cut it off while chopping wood for the family stove at a young age. The whole family wore thongs and he adapted by having his second toe in the thong between his second and third toes.

We all loved lying in the truck on our parents' double bed. This was the prime viewing spot and we fought over it all the time. We did not believe in dibs. First one in got the spot and quite often us little ones had to share with the bigger ones until we got bored and moved on to playing a game of some sort. If we had an old pack of cards we would play cards or "I spy" or make up silly jokes.

We were camped at our usual spot on the Culgoa Common and we had to go to a job for the APC. Before we left any camp we always cleaned it up – our mess and other people's, in case anyone thought we had left it behind. This particular time, I was in the back of the truck and I had to get down onto the ground. Instead of stepping down onto the empty kero drum that was used as our step, I tried to jump over and past the drum. As I jumped, my long dress got caught on something in the truck and I fell face first onto the drum edge, knocking out all my front teeth. Mum quickly grabbed a clean tea towel and placed it over my mouth to help stem the flow of blood. With the truck all packed up they drove me to hospital where I had several stitches to my gums, bottom lip and down my chin. I clearly remember looking over Dad's shoulder as the doctor and a kindly nurse fussed over me and quietly assessed and reassured me that I would be all right. I spent several days in hospital while the stitches healed.

My parents and siblings went off on the droving trip but not before visiting their friends Keith and Agnes Brummel. One of the family visited me on a daily basis, bearing gifts of some sort and I soon had a collection of pyjamas, singlets and panties. How spoilt I felt, and how grateful. After I was let out of hospital I went and stayed with Mr and Mrs Brummel and their family: Don, Lynette and Shirley in Dirranbandi.

Mr Brummell was a returned soldier from the Second World War and he was bed ridden most of the time. His legs were partly paralysed and he was dreadfully white, painfully thin and of course had other problems as well. Their children were quite a bit older than I was. They had a large vegetable garden out the back and chooks and dogs and a fluffy white cat I could play with. I stayed with the Brummels for several weeks. The worry was I would break open the stitches and living in dirt as we did would not bode well for me. With the flies, horse and sheep poo, heaps of dust, not to speak of dogs being more than happy to jump up and give you a lick or three as a greeting, it was safer for me to stay where I was with the Brummels. I quickly learnt it was a nice way to live!

I had my own bed, three nice meals a day and if there was pumpkin on the plate I was not forced to eat it like Mum made me do. I was treated how I thought royalty would be treated and they were a very kind and loving family. They loved paddy melon jam and on Mr Brummel's good days someone would carry him out to their car and with his wife and me, we would go tripping down the road looking for paddy melons. Mrs Brummel would take a picnic for us. Although I was excited to get back with my family, I was a spoilt brat by the time I returned to the bush. I missed the luxury and the spoiling of the Brummels, but that was quickly knocked out of me.

The Brummel's son Don went on a couple of droving trips with us and one day Dad bought a horse in Dirranbandi and it had to be taken some miles out to the camp. The distance was too far to lead the horse with the reins out the side window of the truck, so Dad asked Don to help him load the horse into the back of the truck. Dad backed the truck into a deep table drain so the horse did not have far to jump. Then Dad hopped into the back of the truck, holding the reins in his hands, and tried to pull the horse in. Don was supposed to gee him up from behind, and this was not too successful so they swapped places. Don was in the back of the truck and Dad was outside with a rope around the horse's backside and heaving on it, the horse was shying back and trying to escape and rearing a little. Don could not quite get the knack of how to pull the horse in with brute force so Dad yelled to Emmie, 'Hit the bastard on the arse.' She grabbed the loose end of rope and gave the horse a huge whack on the bum. It smartly jumped onto the back of the truck and Don just as smartly jumped up onto my parents' bed, away from the horse. A great feat of agility for a young lad.

While in this district, we all got "sandy blight" or conjunctivitis as it was really called. It was dreadful and we kept catching it off each other. All that yucky muck in our eyes and when we woke in the morning, we couldn't see a thing until our eyes were washed out with warm salty water. We got this on and off for a while and eventually grew out of it. Another problem was the flies stung our eyes and we would end up with a bung eye. Flies are a dreadful nuisance and the biting flies did not have to be near you for long before they stung. This was the beginning of the "Great Australian Wave", trying to stop the pesky bastards from landing on our face. We learnt how to squint and pull faces or to shoo them off before they could sting. If you were carrying a handful of wood or steel pegs for the sheep break, you had no free hands to shoo the flies away. There was no *Aerogard* in those days.

One time we had a mob of sheep on the road between Dirranbandi and St. George and it was time for fresh meat. Dad was skinning a sheep and the St. George policeman drove past in his car, did a U-turn and came back. As the cop drove his vehicle over the tough black soil and large tussocks of dry grass, Dad quickly slashed the ears of the dead sheep's head and threw them to Gus our dog who was panting nearby. The ears had the station owner's branding on them so that made short work of getting rid of the evidence. Dad then cut the brand off the skin and stuck it up the sheep's backside. Dad rolled the carcass back on top of the wool and casually lit a cigarette, waiting coolly to greet the cop.

When the cop pulled up, Dad had a chat with him and asked, 'Do you want to wait around? You can have half the sheep.' He continued to dress the sheep but the skinning knife was a bit blunt and as he walked off to sharpen his knife, the cop asked him what he was doing. Dad told him and the cop said, 'Here, have my pocket knife.'

Obviously, he had no idea how to skin a sheep. After the sheep was cut in half and then quarters, the cop assured Dad he could cut the chops etc. so Mum found an old sheet that was kept for wrapping the fresh meat in. Dad wrapped it up and said to the cop, 'It's a bit hot mate.'

The cop replied, 'It will be cool by the time I get back to the station.'

This always tickled our fancy as we knew what the double talk meant: "hot" meat was code for stolen.

Sometime after this the policeman had his young daughter with him. We were giving her a ride on one of the ponies and she was highly delighted with this treat. The cop asked Dad if he had any strays and Dad said, 'Yes, but we will be putting it out of the mob tonight.' He said this to let the cop know he did not condone stray sheep in his mob.

The cop surprised him by saying, 'Why do you want to do that?'

Dad promptly caught the offending sheep, tied its four feet up so it could not jump around and plopped it into the boot of the car and away they went. Lamb chops for dinner!

A year or so later, Dad decided that Mum should get her driver's licence and Dad drove into Dirranbandi and pulled up opposite the police station. The cop looked up from doing his paperwork and said, 'What can I do for you, Mick?'

'The Missus wants to get her driver's licence,' Dad replied.

'But she's been driving for years,' the cop said.

'Yes, I know that but she still wants it!'

'Where is she now?' the cop asked.

'Out in the truck,' said Dad. 'You had better take her for a test.'

All through this, Mum had been in the truck scared stiff. When she came inside, the policeman stood her up against the wall and took note of her full name, height and age. He then copied this information onto another precious piece of paper and handed it to Mum with a flourish. 'All legal now, Mrs Kemp,' he said.

Dad asked the cop what did he owe him and the cop replied, 'That will cost you a sheep, one of those "hot" bastards.'

These policemen became our best friends and often received some "hot" meat. The "good" stock inspectors also received legs of lamb or whatever was available as they never failed to bring our mail and loaves of bread when they passed by.

Being such a large family, Mum and Dad never went anywhere without a loaf of bread and a tin of fig jam, a tin of cream or possibly a nob of devon to help out with any invitation to a meal. Also, it was not uncommon to be asked to come for dinner and bring your own or a portion to add to the dinner being arranged. If we had a "killer" (a sheep to be killed to be eaten) the meat would be shared. Living on the reserve we had to be careful of killing stock in the area. If a killer was needed, Dad and Col would drive a few miles out of town, send a dog to round up a sheep, and they would kill it on the spot, hide the evidence and bring the carcass back. Dad would sometimes drive out to a property where he had worked and they would sell him a sheep. Dad was very protective of his reputation and although he never appeared bothered to steal a sheep, he didn't want to get caught doing it. Most stations, whether sheep or cattle, carried their killers in the horse paddock so they were handy to get when necessary. Station killers were always fat lambs or no older than two tooth (two years old). If the station owner denied Dad a killer, he would go back some time later and help himself to one or two and possibly share it with the local cop. Dad was always generous with the "hot" sheep and he never denied any callers fresh meat. He always said, 'It does not hurt to keep in the bastard's good books,' meaning the local constabulary. Some could get really tough on the drovers for various reasons, in particular the young cops. Dad found them a bit too keen regarding the law. The older, longer serving cops were kinder and had empathy with the drover's and their lot. Most drovers did not push the barrier, being aware how difficult life could be by upsetting the law, and the same applied to the Stock Inspectors.

We were travelling to a job one day and driving through Dirranbandi with the truck loaded up with six horses, kids and the usual gear carried for droving. As Dad drove along a young cop drove straight across the road in front of the truck. With much cursing Dad jammed the brakes on, the horses were stumbling in the back, trying to keep on their feet and Dad managed to stop within an inch of the driver's door. Dad jumped out the cabin and checked that he did not hit the car and shouted to the cop, 'What did you do that for, you stupid bastard?'

The young cop said, 'I wanted to check your brakes.' He spoke like it was the normal thing to do.

Just as well he was in the police car or Dad would have jobbed him, deservedly so.

Dad loved people and when he went visiting, he would drag Mum and us kids with him, except for Col being the minder of the camp. Taking us on the trip made it easier to keep an eye on us as we were mostly kept in the back of the truck and our beds were there so we were set if we went to sleep. One time we were quite a distance out of town and Dad decided he would go and visit Mr Brummel as he had heard he was not too well. We were all loaded up except for Col and off we went in the old rattly truck. It rattled because of all the stuff hanging inside – hobbles, chains and other gear that was necessary for droving.

Mum, Dad and Emmie went inside the Brummel's house. This particular time they were having a long visit and we probably went to sleep and then woke up. We were bored so one of us decided to go to the toilet, the outside dunny behind the house. As toilets were a bit of a novelty, all four of us went: Mary, Les, Mike and I. We were climbing the fruit trees and cuddling the cat and the dog and having a great time. Eventually, when we got back around to the front of the house, the truck was gone – disaster. Were we going to cop it!

When the parents arrived back at the camp, they called for us to get out of the truck but there was no answer. Called again, no answer. They thought we were pretending to be asleep and Dad got up into the truck and realised all the beds were empty, so they had to make the long trip back to Dirranbandi and pick up a mob of sheepish, scared kids. Thankfully, we never got the flogging we thought was coming and we never did all go to the toilet at once again. I think my parents learnt a lesson too because they always checked we were in the back before driving off!

Also living in Dirranbandi was a Hawker, an Indian called George Box who had a brother in the fruit and vegetable business. They had a shop in town and also an old truck. Once a week Mr Box would fill the truck with goods from the shop and drive to surrounding districts selling his wares to station owners and drovers as he came across them.

He had boxes of fruit and vegetables, chewing gum, lollies and odds and ends such as sewing cotton, torches and other necessities. He would also barter for old wool, sheep and cattle hides that would be left to dry out hanging over fences. We loved to see Mr Box driving to meet us on the side of the road. The first thing he did was cut apples in half to share amongst us kids and then he would do business with Mum. One day as he was driving away from the camp, a box full of oranges fell off, the wooden box disintegrated and we kids ran madly behind his truck shouting, 'Mr Box, Mr Box, a box has fallen off.' He heard us after a screech or two and pulled up. We helped him gather the oranges and placed them in the truck. He gave us an orange each for helping him. An orange each was like winning the lottery as usually, Mum would cut an orange into quarters and share that among us.

One time we were left in the camp while Dad and Mum went to town to socialise and look for work. Col had a brain wave, and we cut the dry grass into a huge pile and made an emu nest out of it. We sat in the middle and sucked on cigarettes that Col had stolen out of our parents' supply. They were the "roll your own" kind and we all had a puff but Col and Les got right into it. When the parents got back to the camp, Les told on us all for smoking and we all got a hiding, except for him as he had told. He was a "brave boy" for dobbing, which he did on a regular basis.

Dad was a red head and Mum a brunette and though we were all born blonde, Col and I grew into brunettes while the others grew into red heads or "carrot tops" as they were often called. Les was the only one to be born a red head and he was favoured from the start. Even at a young age, he was a cruel ratbag to us "young ones". If we complained about him pinching us, rough teasing or stealing any treat we had, he would deny it and the parents would believe him so we never had a case.

Across the river from our camp. the Aboriginals had their humpies and campsites. We could shout out to them to say hello but mostly we got insults back. It was always done in fun, the black kids having a way with words. Quite often their parents would have a go too. At the end there would be a wave and lots of laughter. If the Aboriginals were serious they could easily swim over and sort us out but they were as lonely as we were for a bit of chiacking around. If they tried to get us going when the parents were home we did not take them up on it, as much as we would have loved to. They would yell out to us, 'You yella dogs, go on 'ave a go, go on, scared of your mummy and daddy.' We would just walk around doing our thing, managing to ignore them. But when our parents drove out of the camp, before the dust had barely settled it was on again. We would chant back to them, 'Sticks and stones will break our bones but names will never hurt us.' Normally we would be in our "attitude" stance, hands on hips and poking out our tongues and a few rude hand signals.

When we went into town we were never allowed out of the truck so eventually a lot of the Aboriginal kids would come up and stand at the back of the truck for a chat. They would have a bag of hot potato chips, lollies or chewing gum and tease us by offering a piece and not handing it over. Sometimes we might barter something that was in the truck but we could never let Mum see we had taken a treat off the other kids as that would have been deserving of a swat of the hand on any part of our anatomy within reach.

In 1956 we went to a property called *Trafalgar* (next door to Cubbie Station, out of Dirranbandi) to pick up a mob of sheep and the station manager was Bill and Alice Holmes. We had to muster the sheep before taking them, so we spent a week or more there while the men mustered. Spending time like this on a property was always like a holiday, not having to pack and unpack every day and quite often they let us stay at the shearers' huts so that meant we had showers also. This made a nice change to bathing in the big round tub once a week, whether we needed it or not.

The Holmes were a family of five, two daughters Shannon and Cleone and son Bryon. Emmie became very close to the family and she and Shannon were pen friends for years. Emmie and Bryon married ten years after they met, when Emmie was eighteen. Byron used to joke that when he met us we used to run around with no pants on and were really dirty, much to Emmie's embarrassment.

Over the years we met all kinds of folk, good and bad. One particular person who was a bit different to others was Gordon and he owned a station between Dirranbandi and Nindigully. When he was younger, he had been trying to knock down a tree with a bulldozer. It would not give, so he went at it full pelt and a limb fell on top of his head and tragically this affected him badly. After he recovered, he was never quite right in the head. For example, he was sick of the perfectly good house on his property so he bulldozed that one down and had another larger house built that he thought was more suitable to his station in life.

One weekend all the station hands and Gordon went to a circus in St. George. They walked around the site admiring all the wild animals that were tied up or in cages. One cage held an old scraggily, ill kept lion and Gordon walked up to it to give it a pat, the lion gave a lazy half-hearted roar, snarled at him and lay down on his belly. One of the circus workers walked past and Gordon engaged him in conversation with an unusual outcome. When the circus was over, the men had the lion in the back of the truck and home they went. The lion spent the rest of the night in the back of the truck and the next day they proceeded to build a large yard for the "new pet". After a few months, Gordon had him walking on a chain like a dog. He had hired three Aboriginal families to do stick picking on the property, so there along a dusty track would walk Gordon, the lion, a collection of his dogs, and the Aboriginal family all enjoying an outing in the late afternoon.

Gordon and his workmen went to church on an irregular basis in Dirranbandi on a rough dirt road. He later decided that he and the men needed to go to church more regularly so he decided to build one on his station. Gordon would get dressed up as a minister in a black robe, full black gown with all the regalia and take the service. On occasion, he would strap a colt .45 on his hip as well and the men would say they did not know whether the boss was going to be Jesus Christ or Ned Kelly for the day. Sometimes after the church service, Gordon would mount his horse and gallop off, his robe flapping in the wind, shooting the gun in the air.

Later on, he lost all his money and the bank took over his property. The day he left he set fire to the house, burning it to the ground and drove the bulldozer into the dam.

WOLLONGONG 1957

Traditional land of the Dharwal or Turulwal people (co named)

Beryl Junior was born in July 1956 in Narrabri New South Wales. She was to be the last of the family bringing us to seven kids and two adults. It quickly became Emmie's job to look after the baby and pack the camp gear up so that freed Mum to put the sheep break up and down as we could not do the job ourselves. If Emmie finished before we did, she would help carry the pegs but if Beryl was up and about, the baby was her sole concern.

By 1957 Dad was sick of the stress of never getting ahead and always trying to make a quid. Money was hard to come by and some people never paid him or only paid a portion of what the job was worth, so he decided to move to Illawarra. He sold his droving plant and took the family to the city. Dad's idea was to eventually buy a butcher shop. He planned to camp in the backyard of his sister Anne's house, where his mother Nanna Thomas also lived, and work at the smelters.

While in Lake Illawarra, both Mike and Beryl got asthma for the first time. Mike also got the mumps. Beryl, not to be outdone, drank creosote that was stored in a bottle out in Uncle Wally's shed and that meant a rushed trip to hospital to have her stomach pumped. She was in a bad way for a while, not expected to live. Showing true Kemp spirit, she rallied and had a healthy life like the rest of us.

One day Les had the bright idea of surprising Nanna Thomas. He told Mary and me to crawl with him under her bed. Eventually, Nanna came into her bedroom and sat on the edge of the bed. Les grabbed one of her ankles real hard and pulled. Nanna Thomas let out an awful scream and so did Mary and I. This loud noise had to be investigated of course and in marched Mum and Aunty Anne and we three got a good walloping. It was probably Les' very first one! Mary and I never pulled that trick again but Les did. In his eyes it was quite funny so that made it all okay. Mum and Nanna Thomas were never friends and Mum feared and hated her with a vengeance so that was possibly why Les decided to do what he did. Mum often made comments about Nanna in front of us kids, so maybe Les thought of it as a bit of a payback for her.

We all went to school while in Lake Illawarra. It was my first time in any school and I found it very distressing. I was only five and I cried and cried for my brother, so I was put in the same class as Col, aged ten. I sat beside him and sobbed and drew pictures all day to amuse myself. I was away from my sisters and did not like it one bit. Col was very embarrassed with my presence there and this stint of schooling was the total amount of Col's formal learning. The bigger boys pushed me around a bit so I then went and sat with Mary aged eight in her class and that was much better, not that I learnt much. This was the only time Col, Emmie, Mike and I were in a school room to learn.

While living there we had a bit of a social life. Aunt Anne had three kids, Fred, Wendy Pout, and their half-sister Juanita who was Emmie's age. Also living in the house was a young man, Tom Elford, who worked for Uncle Wally. Tom and Wendy got married later on. Our cousins were allowed to go to the pictures each Saturday afternoon, so we were allowed to go too. This was a real treat for us, going to the pictures on a regular basis and being able to see if Tarzan got out of his predicament from week to week! Unfortunately, the very first time we went, the crowd of kids was so thick going out at interval and me being so short, I did not get a pass out. We had an ice cream and when it was time to go back in again and I was asked for my ticket. It was a case of: 'What ticket?' I was a country kid and this was my first matinee. What confusion! I was with Col and he cottoned on straight away that I would not have picked one up on the way out. I stood with my head down and started to cry, sucking madly on my thumb, thinking I was in a great deal of strife. Col told them that a few bigger kids had been hassling me and maybe they had managed to get it off me. Reluctantly, they let me back in, but I never forgot that experience. The next week I asked for it as I went **in** as I did not want that to happen again. Every Saturday we had to go back to see Tarzan's cliff-hanger – would he survive?

Sometimes we were not allowed to go, maybe one of us was being punished for a misdeed, or maybe there was not enough money to let us go again. I think our parents liked us to go just to get us out of their hair for a while. The house was full with ten kids and five adults living there. We were camped out the back as if we were droving, though Emmie shared a bed with Wendy, Mary bunked with Juanita, and Col slept on the floor in Fred's room.

On the odd weekend, Uncle Wally would bundle up all the kids and take us to the beach for a longed for dip. He taught us to dig our toes in the sand, twist our feet down until we felt pippies, which we then picked out of the sand and put in a bucket of water. When we arrived back at the house, Aunty Anne prepared and cooked them and we all had our first taste of seafood, though I have been told the yabbies we got out of the bore-drains were sea food! We all liked them as it was a bit of variety from meat, potatoes and bread. Before going back to the house, Uncle Wally would give us all an ice cream. He was a tall, well-built man and he would hoist us up onto his shoulders and carry us to wherever he wished to plonk us. He was a lot more fun than the other adults.

Television had now arrived in the city and some neighbours up the road had their TV in their front lounge. As a group we would trail up and sit on their two foot high, brick front fence and watch it. We could not hear it of course and eventually one or the other of the kids would start to giggle or fight and after a while the TV owner would get sick of the noise we made and the carry on and pull the curtain on us. At this stage Aunty Anne did not have a TV, but while we were there they bought a small one. This TV had rabbit ears and to watch it we had to feed it pennies at a time. Then there was always a fight between the kids over who was going to put the pennies in to see the continuation of whatever show we were watching. The adults always watched the news but we got to watch a small amount of other shows on it. I always wondered how they got the little people in the tiny boxes and when no one was around I would look behind the TV to see if I could catch them. I never did manage to work out how they got in there!

Dad was offered a job at Port Kembla Welding Works and he stayed there for several months while he learnt the trade. The butcher shop did not eventuate and the welding work did not suit him either as it was very hot work and Dad developed dermatitis there. It was mostly between his toes and fingers and this stayed with him for the rest of his life. He decided living in the city did not suit him at all so he bought an aqua coloured Austin truck and we went to Moruya to visit his brother George and Aunt Betty.

Aunty Betty was an Aborigine and they had no children themselves but adopted a girl called Joan who was disabled and in a wheelchair. Joan would have been in her early twenties and did not speak very well. This was the first person we ever met who was in a wheel-chair and we were a bit in awe of her. When we got to Moruya, Mike got mumps again, on the opposite side of his face and was very sick.

We kids adored our aunty. She was always ready with a hug and a kind word, which was sadly lacking in our family but we did not like her cooking at all. She had peas with all meals and they were the dried type that were cooked with soda and tasted vile. So when we had a meal, we would sit on the veranda, making sure we sat near a crack in the floor and when no one was looking we would push the peas through, as we were forced to eat everything that was on our plates.

Uncle George collected old bottles and recycled them. He had long brushes and lots of soap and I loved helping him, kneeling beside the tub Emmie, Mary and I washed and cleaned until we could wash and clean no more. What little girls did not like to play with water?

While we were camped in their backyard, Dad went pea and bean picking, but he longed for the bush so before long, he packed us up again and off we went. Dad had to take his mother to Dubbo first before we moved back to Narrabri so her goods and chattels, along with our own, were loaded on the back of the truck and we were plonked on top. Beryl and Mike were in the front with our parents and Nanna, and away we flew. Dad certainly drove as fast as the truck would go and much faster than the law allowed. We were perched on top of the load with nothing to hang onto and thought it very scary as the view whizzed by with the wind making our eyes water if we dared open them. Crossing the Blue Mountains there was a big bridge and we could see the water in the river way below us. The next moment we heard clatter and bang as Nanna's pee pot went flying in the wind and disappeared over the bridge and out of sight. We looked at each other and just hung on. Eventually, when Dad pulled up to check on the load and us five kids, Col told him about Nanna's pot and Dad went a bit crook at him for not yelling out telling him to stop the truck.

I heard cicadas for the first time on that trip. I did not know what the sound was but I could hear them clearly over the sound of the truck and the wind whistling past. I didn't see one until some forty years later.

We spent the first night in Wellington at an old friend of Dad's, Bill Boland. The next day we continued on to Dubbo where Nanna and her goods were dropped off at her son Fred's house. We saw our cousin Shirley who was around Emmie's age and we stayed there for a few days catching up on the family gossip. Then we continued to Narrabri.

Dad camped once again near the McCaws. Out came the trusty

tent and it was set up in our designated camp area where we settled down to wait for the next job to come along. Across the road from where the McCaws lived, was the local butcher's kill yards that operated about three times a week. Mrs McCaw would take an old pram over to pick up meat for themselves and bones for their dogs. We kids took over this job and we always managed to get enough for both camps.

We had heaps of dogs and so did the McCaws. One time Les, Mike and I were walking to the slaughter yards and Les decided to stir up a bee hive in an old gum tree that we had to walk under. He was a good shot with his trusty homemade slingshot and we were soon under attack by a great swarm of bees. Les outran them, but Mike and I, being younger and puny, were stung by quite a few angry bees, mainly around our neck, head and arms. Mrs McCaw was astounded that they had attacked us as they never bothered her – but she didn't know Les very well.

There were plenty of ducks and chooks and a hen had set on a few duck eggs. The ducklings hatched and mother hen fussed and clucked over "her" babies. The hen took the ducklings to the pond for a drink and the first thing they did was jump into the water and swim around. The hen put up a dreadful fuss, trying to get them out of the water. She ran around the pond, clucking madly, flapping her wings, frustrated that they would not obey her. As they paddled across the pond she ran around to them and then they would paddle on to the other side and she would race around to meet them there, cackling and fussing. We thought it was a great show. Eventually they got out of the water and the hen took them away into the shade to have a feed and a rest.

Mum's nephew, Tom Fields, had married Mrs McCaw's daughter Lucy. He was out of work and Dad managed to get a couple of small jobs and he invited Tom and his pregnant wife to tag along on the trip and do any odd jobs they could find. They had a Land Rover and like Dad, not much gear to start with. When they joined us on the road helping Dad with the small mob of sheep he had, Tom and Lucy spent a lot of time in paddocks picking up dead wool – this is the wool off the ground that is from dead sheep. They had a forked stick that they used as a comb to get the wool off the carcass. When they had filled a chaff bag or 20 pound potato bag they took it into town to cash it in. Sometimes there would be big clumps of wool on the ground that had fallen off sheep that had not been shorn for a while. Dad had collected wool like this at times when he was hungry for a quid too.

Dad never went past a piece of copper, brass or old batteries. These were during the times when jobs were hard to get and Dad continued collecting things until he could demand a decent wage for his hard work, which eventually he did get. These collectables were put in a 44 gallon petrol drum with the top cut out and when he had collected enough, he would sell them. You may wonder how we all fit into the back of the truck – I can tell you – with great difficulty.

After we returned to Narrabri, Dad upgraded his vehicle to a truck with a base about 20 feet long and he decided to put a permanent cover on the truck. Up to this point in time, the truck only had a steel frame that was covered over with a tarp. Dad mated up with a steel works in Narrabri and did his own welding. In doing so, he missed quite a few spots on the roof of the truck and he had to eventually put tar across the joins so everything did not get wet in the event of rain. This crate was transferred to each new vehicle he bought over the years. It was made with an added steel frame that also hung over the cabin of the truck. When it was finished the whole thing had to be lifted onto the back of the truck with a crane. This increased the overall weight of the vehicle a lot. The bottom half was hard pressed wood, and the top half had a tarp that could be rolled up to let the cool air in when needed. This meant of course the truck stayed open for weeks on end unless a storm brewed and then the tarp could be quickly let down. It was still very cold sleeping there in the winter.

Mary, Mike and me, wore the same clothes in summer and winter. For extra warmth in winter we would put on a jumper or a cardigan and when it was really cold, we would rug up with a cardigan over our jumper. Mike wore shorts and Mary and I wore a dress or skirt. Beryl always had long pants of some sort and along with the stockmen, got to wear shoes and socks. If any shoes were bought for the rest of us it was white tennis shoes or the plastic ones that were available then.

We called them "gins" as only "poor" people wore shoes like that. If we had shoes that were no longer "good" or we were growing out of, they would become our camp shoes, which was a blessing as it was mighty cold on those frosty mornings with bare feet. Socks were unheard of except for Dad and Beryl – they always had a pair and Mum would wear a pair of Dad's socks in the chilly weather.

Our shoes could not be handed down, as we went boy, girl, boy, girl and the sizes did not fit the next in line. Mary managed to share some with Emmie but she was well built compared to me being the "skinny ring" as the family so fondly called me. Nothing much of hers was handed down to me and I was five years older than Beryl so she did not get mine. Mind you, if Mum could make one of Mary's dresses fit me, I had to wear it. We had so few new clothes and not many second hand came our way, so we pretty much wore our clothing out. Our life in "the long paddock" entailed a lot of climbing through fences and barbed wire was murder on clothing. We would get snagged in anything that was sharp and dangling. And of course, stick picking for the family fire was one of our jobs. As we dropped the sticks beside the fire we would quite often hear "rip" as our clothes were snagged by one of the errant sticks. Mary and I seemed to always have the middle of our dresses torn and hanging down, Mum would sew them back up for us.

I was now getting old enough to realise the rocking of the truck, along with the rattling of the gear hanging off the walls, was not caused by strong wind outside but the movement of our parents having sex. I was not quite sure what this involved, but I knew enough to keep quiet and pretend I was asleep as the other kids did. If one of us inadvertently let out a loud yawn, Mum would say, 'Who was that?' Not a word would be spoken by any of us and we would lay very still. It was not unusual for her to smack us for being awake... yet we were awake because they had woken us! This was a reoccurring problem we had, as they had a very regular sex life that was prolonged long into the night, at least that was how it felt as we lay in bed, trying to be very quiet and still and praying for sleep to reclaim us. After sex, they had a cigarette each and a chat and we could listen and catch up on some gossip, but I think we mostly fell asleep to the drone of their voices.

One day on the way to our camping spot we visited the McCaw family for a cuppa. When it was time to go, we were in the truck saying goodbye and Dad told Col to close the sliding door at the back. Col quickly slid it closed, not realising I had my left hand around the edge of the truck. I had to have many stitches on both fingers. The tip of the ring finger was taken off completely and my middle finger was squashed. I still have two deformed fingernails on my left hand. When the nail grows on the ring finger, it curls back much like a parrot's beak. It was very painful and it took ages to heal but eventually it did. Mary and I were both left-handed and we were occasionally called "cacky handed" and "south paw". If we were boxing with the boys Dad would yell, 'Hit him with your left paw,' as we had a chance to get a hit in if we were fast enough. It was extra sweet to see the boys bend over as if in pain but most of the time they would slap wherever we hit and pretend a mosquito had bitten them.

Sadly, Alec McCaw was drowned trying to save a mate from a swollen river a couple of years later and not long after, John McCaw died from cirrhosis of the liver. Mrs McCaw was hit and killed by a drunk driver when she was walking into town one day. Her brother and sister, Norman and Amelia, packed up and moved into Narrabri, preferring that to living on the reserve as they were now quite elderly and found it difficult on their own in the rough living conditions.

CHARLEVILLE 1957

Traditional land of the Mandandanji people

Beryl was now nearly two years old and Dad and Mum decided it was time to upgrade to a caravan. Dad came home one day towing a lovely brand new aluminium "Caravel" caravan. After us living in the back of the truck for years we thought it was a mansion. We now had new sleeping arrangements. The van had a double bed at the back and another double bed could be made up at the front where the table was. Beryl shared the bed with Dad and Mum while Emmie, Mary, Mike and I shared the double bed at the back. Les chose to sleep in the back of the truck, having a single bed to himself and Col still camped outside on his stretcher.

The caravan had an ice box fridge. An ice block about 6x12 inches (15x30 cm) was placed in the top of it and lasted about three to four days if you did not open the door too often. How lovely it was to have butter that was firm to spread on our bread and the cooked meat stayed fresh for longer. Only the larger towns had ice works though so when we were on the road and miles out of town we did not have the pleasure of a cold fridge. The fridge made a good meat safe with the meat wrapped in tea towels or it became a cupboard until we could get more ice.

After getting the caravan we did away with the tucker box. It became our winter clothes box and the utensils were now kept in the caravan. This meant Emmie, Mary and I made several trips each meal to set up the condiments and utensils needed for that meal or a table or blanket outside. Although we tried to get everything we needed there was always something missing. Dad could not have a meal with his PMU hot sauce or the syrup would be forgotten or someone wanted treacle instead of honey. If we weren't moving on, everything could stay on the table outside for the next meal. Only opened tins of jam, butter and bread had to be kept inside. The horses loved a loaf of bread and if they could get near the dinner table they would snatch it up, no sliced bread in those days.

The van was treated with great respect. It was the first new vehicle Dad had bought. One day, Mum was leaving the campsite and she got into a bit of difficulty reversing the truck while towing the van. It went into a sharp turn and hit the back gate of the truck, putting a one inch hole in the top left hand corner of the van. Mum was dreadfully upset. When she confessed to Dad he hit the roof. He ranted and raved about how she should have been able to reverse and how she should have been more careful. Dad quite often referred to that incident not letting anyone forget about it.

We hadn't had the van for long when one day another drover stopped to say hello. Mum went out to see who it was and was soon followed by us. As was the custom, I closed the door behind me to keep out the flies but little did I know the keys were inside. We had to cut the gauze on one of the windows and Mike was lifted through to open the door. Another little "incident" we didn't forget in a hurry.

To our standards, the van was pretty swish. Inside the main door was the gauze door. The doors could be unclipped from one another so the main door could be hooked back onto the side of the van leaving the fly proof one closed. What a treat this was when we did eat dinner inside – we didn't have to share our meal with so many flies. This of course did not mean that no flies came into the van. They actually stuck to the back of our shirts, which meant much waving and slapping to get rid of those that found their way inside.

Jack Smythe asked Dad if he would take on the job of moving 1,600 head of unhandled, wild steers from Mungindi via St George and Mitchell to the Charleville saleyards. From there the steers would be trucked further up north to another cattle station Jack owned. We did not do many trips with cattle so it was a very exciting time for us kids. Working with cattle meant less work for us, as our biggest job was to collect the wood for the fire and to look after the dogs.

A trip like this had to be fully prepared for from the start with men and provisions. Dad hired three stockmen, counting on my brother Col as the fourth. Col was eleven, tall and well built and the only one of us kids who was capable enough to ride with the stock. Mum frantically began washing and patching clothes, baking fruit cakes and brownies, and working out what groceries to buy to last two weeks or more as there was such a distance between towns and the likelihood that we could be stranded by wet weather at any time.

Fresh meat was bought to last the first few days: sausages, cold preserved meats and other cheap cuts such as Devon. We would use this for lunch (dinner time) but at night (tea time) there was always a cooked meal. Twelve loaves of bread were ordered from the baker to begin with and after this was used, dampers and scones were made in camp ovens in the coals. Two dampers only lasted for one meal, bread and meat was the main staple in any drover's camp. Mum would give us dried Weetbix with syrup, honey or Vegemite for dinner and the dampers were kept for the men. This saved a lot of baking for her. We didn't ever whinge or quibble over this sort of meal. We thought it was quite acceptable unless one of the non-workers had a slice of bread and then you would hear: 'That's not fair.' We could always choose to have the cereal with powdered milk if we wanted. Butter was a real treat on the road as we had no way of keeping it cool enough not to melt. For a while we tried wrapping it in a wet tea towel and placing it in a dish to stay damp but once it became very hot there was no way of keeping it cool. There weren't many towns to get ice on this trip.

Dad had to decide which of the horses were more suitable to take on the cattle drive as only six would fit in the back of the truck at one time. The remaining horses were left on a friend's nearby property or on a station for agistment for the duration of the trip. Along with the horses we had the droving gear and the workmen's gear. The men brought along their bedrolls and a suitcase or an old army duffle bag to fit their clothes in and it was up to the drover to supply the working gear and saddles. Cold and wet weather gear, which was usually a waterproof bushman's raincoat or jacket, was tied behind the saddles.

It was not unusual for Dad or Col to share their work clothes or work gear with the hired workers. Upon arrival at the next town, Dad would pay wages and encourage the men to get appropriate working gear to finish off the trip. Dad was a hard boss and not all men who started the trip finished it. Dad would leave them on the side of the road with their swag and let them hitch a ride to the nearest town. If we were a week or two into the job, the stock would usually have settled down to a certain degree so Dad could get away with one less worker and of course, a worker less meant a wage saved!

Col became the horse tailer on this trip. Every drover had a horse tailer whether they had four horses or twenty, a horse tailer was a must. At the end of each day he would take the horses to water and secure them on grass with hobbles and place bells on the flighty ones. He would have a night horse that was trained to be tied to a tree on a long chain called a tether and that horse could nibble on any grass within its reach. A night horse was a special horse, trained to carry anyone on their back, easy to ride and handle at the most difficult of times. If grass was not plentiful, the horse tailer would feed the night horse a bucket of chaff and oats. It was Col's job to put a rug on the night horse. In the morning it was his job to carry the rug and the chain back to camp. In the early days, sometimes Col would forget to put the chain in the back of the truck and we would drive off without it and it would not be discovered until we landed at night camp. Sometimes Dad would drive back for it but at other times he would make Col canter back on his horse and retrieve it. That was his punishment for forgetting it. It only happened a few times as we kids made it our business to look for it or Col made sure it was dragged closer to the camp where we would find it. The chain was about twenty feet long and the only way he could carry it was to wind it back around the horses' neck. The horse tailer got up about an hour before daylight to find the horses, take their hobbles off, tie them around their necks and then bring them back to camp, ready to be caught and saddled up by their riders.

We set out in the truck, six horses, nine dogs, three stockmen, our family of nine and the caravan. In the front of the truck with Dad was Mum, Beryl Jr and our old friend Alan Alston. In the back the double bed stayed down and this is what we kids sat or laid on. The other workers sat in the back with us. There was an older man called Hughie, an Aboriginal who Dad had to drive fifty miles to pick up but later said was not worth driving fifty yards for; and Alf Adare who had worked with us a few times before. To fit six horses in, along with all the food and gear, the two single beds at each side of the truck were tied up to lie flat against the wall. As you can imagine, there was not a great amount of space!

When Dad stopped for a meal break or for any length of time, we had to climb over the horses' backs to get out of the truck. Most of the horses did not mind but some would try and pig root, meaning they would drop their heads, hump their backs and do mini buck jumping or rear up. Sometimes we crab crawled along the side of the truck over the horses' heads and tails and once we got to the back door, we slid out between the rung and walked down backwards, much like a ladder.

Dad was given a mud map to find the station we were heading for, which was situated in the wilds between Dirranbandi and Mungindi. The stock were to be mustered and then taken on the stock route to Charleville, where they would be loaded onto road trains and trucked up north to another station Jack Smythe owned.

After driving for several hours, Dad realised he was driving in circles and the sun was going down. In the headlights we could see kangaroos by the dozen jumping in front of the truck and racing along the road in front of us. There were also bunnies with their white tails bobbing, foxes with long bushy tails, dingoes scurrying to get off the road and the occasional emu with their long, even gait, keeping up with the truck for a while before veering off into the darkness.

Eventually, Dad admitted defeat. We were lost. He decided to go into a property that he could see the lights of in the distance. Up the rough track he drove and an elderly chap walked out to meet us, carrying a lantern. Dad told him where he was heading and the old man tried to explain to Dad where he had to go but the directions got a bit confused. The old chap put his arm around Dad's shoulder, turned him around and pointed his other hand out over the sky.

'See that big star over that tree there? Well, right under that star is where you are heading.'

Dad thanked him and, trying not to laugh after being given these useless instructions, stepped back into the cabin of the truck.

After travelling a few more miles up the road Dad decided to stop and rest for the night. It was pitch black so he parked on the side of the road. He could not go too far off the road as it had been raining for a couple of days before and we were in black soil. The table drains were running with water. Everyone in the back of the truck climbed over the horses, grateful to get out and stretch our legs. The billy was filled with water from the running table drain and set on the gas stove in the van.

As the billy boiled, the gear was taken and placed on the ground. The dogs were also taken from their crates and tied around the truck, after being given a drink from the table drain. Tea was made and we had a sandwich and then slept where we could find a bed as the horses stayed in the truck. We younger kids and our parents slept in the van, Alan Alston slept in the front of the truck, Col managed to pull his swag out and lay under the truck, and the other workmen along with Les slept in the back of the truck.

The next morning Dad woke us at dawn and told Col to fill the billy from the table drain. As Col knelt down he saw a dead sheep in the drain about ten feet up from where we had dipped the billy the night before. Needless to say, we used the water from the drum on the side of the truck for our breakfast cuppa and milk for our cereal.

None of us fancied a "drink of dead sheep" for breakfast!

Dad was driving along at a brisk pace when Col shouted out, 'Stop, stop.' The caravan had jumped off the tow bar and was wobbling along behind, being towed by the safety chain only. Dad jammed on the brakes and the truck came to a stop with a shuddering halt. The horses wobbled all over the place, trying to keep on their feet. They could not fall over as they had a rail between them and were face to rump to stop them from biting. As the truck stopped, the caravan ran straight into the tow bar, and the van had a one foot hole in the aluminium.

When Dad saw the big hole in the van he started to laugh and Alan had a good belly laugh too. Mum did not see the funny side and completely lost her temper and cursed them for their stupidity. Dad re-hooked the van onto the tow bar, adjusted the safety chain and off he drove, black dust and dirt blowing everywhere. He told Col in future to shout out if there was a problem and he would not jam on the brakes as hard as there was no emergency that deserved a "Stop, stop" shout!

When Dad came to the next turnoff, he realised where he had gone wrong the day before and we arrived at the station where we were supposed to be the previous afternoon. Dad explained to the station manager what had happened. The manager thought we had got bogged somewhere. Dad was informed there were two mobs of steers to be taken to Charleville, shared with two droving plants working independently. The other drover was called Fred Crowther who had four men with him and his plant consisted of pack horses.

Dad unhooked the caravan where it would be parked for the duration of the stay. He then drove to the stock loading ramp and unloaded the horses. The men took them off to have a drink in a trough while he organised a nose bag of chaff and oats for the horses to have a snack. While the horses were eating, Dad and the men had their smoko. Then they saddled up and rode off to meet up with Fred Crowther and help with the muster, leaving Mum and us kids to unpack the truck.

It was wild scrubby country, most of it being mulga and box gum trees, and the men soon disappeared into the bushes. They carried in their saddle bags a hastily packed dinner: sandwiches prepared by

Mum and Emmie, a piece of fruit cake and maybe a tin of fruit was included, which they could open and eat with their pocket knives. The juice was a grateful sweet drink to finish off their midday meal. They also took a quart pot with dry tea and sugar. The quart pot had its own deep lid, which then became the drinking mug. The tea leaves and sugar were placed separately in a square of paper and the top twisted to keep them intact. How much of these products they used, decided how many hot drinks they had during the day. The men never bothered to carry dried milk on camp outs.

Cutlery was not necessary when they had a packed lunch. When in camp, the men would pick up a stick and stir their tea, not bothering with a spoon. Most men smoked and they always had a box of Vesta wax matches in their saddle bag as they never knew when they would need a match to light a fire. When on a muster, morning or afternoon smoko break was not often taken and sometimes they did not get a dinner break, depending where they were with the stock at the time. These stock were unhandled and hard to control. If the men stopped, the cattle would have scattered again so it was best to keep them moving until they could find a spot to control them, maybe in the corner of a fence or some bush yard. The men had to be capable of getting their sandwich out of the saddle bag as they rode along and eat while on horseback. If they needed a pee, they would just pee over the side of the saddle and keep riding. This was also done if a horse was flighty because of the chance of not being able to get back onto the saddle again.

Each of the men had a water bag hanging under his horse's neck and the horses were broken in to carry the water there. Another popular spot to carry the water bag was off the side of the saddle with a piece of hard leather on the side of the bag, between the horse and the bag. Dad said it kept the water cooler and it was up to the men to keep their own water bags full, as no one else would do it for them.

We had never been on this station before and it was very scrubby so Dad told Mum that if they were not back by dark to light a big fire so they would know where to head for home. As the men rode off into the mulga scrub, we looked after the dogs. We then finished unpacking the van and swept the back of the truck and put it all back in order. The two side single beds were lowered back into their respective places and made up.

Col's stretcher was placed outside under a tree. The men's swags and gear were placed in a small pile nearby and Col's bed became a seat during the day as seating of any sort was very scarce in the camp. We did have three wooden stools that were Mum and Emmie's and the first one to grab the other became "the sitter", but as soon as the sitter got up that stool was up for grabs for the next fast person to sit on. We did try to walk around holding the seat to our bums but this was considered "unfair" to the others waiting for a comfy seat. When Dad was in the camp it became his. We were always grateful for a log or a stump that was handy to the campsite and preferably near the fire in winter. We kids never had a seat when the workers or any grown-ups were in the camp – we stood and let them sit down. If we did not do this, we would get a clip around the ear from either Mum or Dad. If they were not close enough to do this they would tell us to: 'Get off your lazy arses.'

The afternoon drew to a close and evening arrived. It was soon dark and the fire was built up and as time went on, Mum started to panic. Great armfuls of wood were added to the fire and we kids shouted and cooeed as much as we could to help the men find their way back to the camp. We had to light a smaller fire for Mum to cook dinner on as we could not get within ten feet of the big one. On the menu was a hearty vegetable-filled curried sausage stew that would keep until the men arrived back at camp.

Much to Mum's relief and ours, the men found their way back. The fire was so huge by this time I am sure it could have been seen from outer space! We were making such a racket Dad was embarrassed and said we were disturbing the dead. I think this was the only time we ever made such a noise without Mum telling us to shut up.

It took a few days to muster enough stock for two mobs to be taken to Charleville. Before starting the trip, they had to be drafted to sort out the smaller stock, as the trip did not warrant taking along stunted or ill looking beasts.

Dad being the boss drover, took the lead with the stock on the road. That way he could keep control of how fast or slow the stock travelled. The cattle were fresh, restless and jumpy and as soon as the steers were let out of the yards they took off at a gallop, trying to get back to their usual paddock. The men took off after them, two riders on each side to try and reach the front. Eventually they managed to turn them with the help of the dogs, stock whips cracking and their trusty horses doing their bit. We took the stock route from Dirranbandi along the Balonne River to St. George, then east along the Maranoa and into Mitchell. We then turned north to Mungallala, Morven and on to Charleville, going through a cattle station called Lesdale about 20 kilometres out of Charleville on the east of the main road, now called the Warrego Highway.

The whole trip was scrubby with mulga, whip stick mulga and various other sorts of trees and small shrubs. This made it hard to see where the stock and the men were, as the stock were so wild the men had to be on their toes at all times, ready for trouble. Later whenever Dad spoke about this trip he would say, 'It was so scrubby, the dogs had to back out of the scrub to bark.' The cattle could not fan out too far as it was too difficult to see if any stock was lagging back or walking out of the mob and the men liked to see the next rider. Mulga is good stock feed so they fed as they walked along. Grass did not grow in the dense mulga but a bit of grass grew in the more open spaces.

The cattle were mongrels for a long while. They would rush at any noise or stick they saw but they never went too far for some reason. All the men had to do their stint of night watch over the stock, which entailed riding around the stock all night, singing or whistling or just talking. They also had to be careful not to get between the camp fire and the stock as the shadows would frightened them and they would gallop off at a whim and they could travel many miles in fear and become very hard to find. The route taken was a normal stock route that went through private property and the paddocks were very large to lose the stock in. The nights were long and the men had two-hour shifts and Col being the horse tailer went on the first shift. Hughie did the second shift and Dad, being boss, always did the midnight shift and then came Alf and Alan last.

That trip was not too good because at night we had to be very, very quiet so as not to frighten the stock with loud or unexpected noises. Eventually, the stock quietened down so all could relax but not too much. We kids went to bed really early to make sure our noise did not startle the cattle. It was difficult for us not to shout or talk loudly. When we needed a pee and had to run behind a tree we had to be extra careful not to trip over a stick or a log and make some sort of racket. Talk about walking on eggshells after dark! Dad realised this and decided to make two fires, the larger fire nearer to the cattle so they had a good glow and then a smaller fire closer to the camp, this was smaller and easier to cook on otherwise there was the danger that as the cook cooked – she'd get cooked!

The night meals were mostly eaten outside and we all ate together at this time. Nine people plus any working men in the one little van was just too squashed. If it was raining, meals were usually cooked outside but eaten inside, with us kids sitting on the double bed at the back of the van. If the rain went on for days, we camp kids, except Emmie and Beryl, lived and ate in the back of the truck, getting out to do the chores when necessary. Whoever was wet usually got to do the all the jobs until they dried out – if they did. There was no sense in us all getting wet as there was no way of drying our clothes and we had very few. When it was wet, we stayed in the same spot for a couple of days so that the truck didn't get bogged in the black soil. The men would put a tarpaulin over the back of the truck for shelter. Some of the gear would be stored under the truck and a fire also could be sheltered there. This was a risky exercise and was not done very often. The risk being that the fire could spread to the truck. The truck would be unhooked from the van and driven a few feet away so we could walk around the vehicles.

Wet weather usually meant a stew or two, with any meat and vegetables that we had, mostly potato, carrot, pumpkin and a tin of peas or beans. To give it some taste, several tablespoons of vegemite were added or sometimes, a packet of soup was used. If the stew was too "runny", it was thickened with flour or rice and quite often we made a batch of scones or damper to soak up the gravy. If it rained for any length of time, we would run out of bread.

Luckily we always had flour, morning tea was often scones fried in fat. We rendered our own dripping off the killers' kidneys. The kidneys were cut into small pieces, put in a small amount of water in a camp oven and boiled until they were rendered into pure fat. We had such an abundance of this, that one of the camp ovens was always half full of fat. We also had a used milk tin filled with fresh rendered fat. Clean fat never appeared to go off and we used it as a spread on our bread and biscuits when required. We got savoury fat out of the camp oven as it had the taste of roast meat – delicious with salt, pepper and Vegemite. We did not like to use new fat for baking potatoes and pumpkin as they never browned to our satisfaction. We would put the eggs in semi-set fat to stop them from getting broken. To get the eggs out, the cook just put their hands into the fat and got the required amount of eggs out of the goo. Dad always had two eggs on unbuttered bread for breakfast with lashings of hot sauce. The eggs were cooked in fat so they had enough grease on them without the added butter and the rest of the family had cereal. Another snack we had if no other spread was available, was tomato sauce on bread or damper with the dripping. We thought this a treat fit for kings. Woe betide us if we used it lavishly, for we would not be allowed to use it for the next few meals.

The truck was always facing away from the camped stock as it gave them a smaller object to hit if they rushed towards the camp and it also gave the men a bit of extra security to sleep in front of the truck. On really cold nights the men chose to sleep near the fire and those on night watch at that time kept the fire well stoked. The wood they used was what we kids had collected upon arriving at the night camp. We always had a heap of wood as it helped keep the sleeping men warm and allowed the night watchers to have several hot drinks during the night.

On the table was tea, coffee, milo, sugar and powdered milk and quite close to the fire was a full billy of water to simmer so that it only had to be placed on the hot coals for a little while to get it boiling. A large tin of Arnott's dry biscuits would also be on the table.

A lot of bushmen who slept on the ground put a rope around their "bed". This supposedly kept snakes from getting into their beds at night as the snakes did not like the feel of the rope as they slithered over it! I'm not sure how well it worked but I don't think I would like to leave it to a rope to keep snakes out! Many a swag upon being unrolled at the next camp, would have a snake slither out and say: "Hi" in passing.

We did not ever have a rush that took the camp thankfully. If that happens lives can be lost, as the stock do not care where they gallop as long as they get away from whatever made them jump. It is very difficult to muster them after they rushed as they can scatter for some distance. Some mornings they would be bunched up together and other times they would be miles away and scattered. When the men heard the cattle rush, they would all jump up and get on their horses tied nearby and gallop after the mob. The person on watch had the best chance of catching up to the lead, as long as he was on his toes to get away quickly.

Dad was really annoyed when the cattle rushed. It was dangerous. The cattle became crazed with fear and the horses took the lead, rarely making allowances for their rider. When they rushed, you could hear the sticks and branches breaking and the hooves crumbling the dirt and stones. About two weeks into the job we had a rush on Dad's watch. Yelling out to the men, he tried to gallop around the crazed steers when his horse went under a low branch and he was pulled off and landed with a whack on his back and was badly winded. The horse cantered on a bit further and then stopped. Dad was unable to explain how or why it happened: maybe a goanna or koala walking nearby, a kangaroo or wallaby jumping along or a branch falling off a tree had startled them. They did not need much of an excuse to turn tail and go.

Sometimes small fires were built around the camped cattle to help stop them from rushing, but this was only done if there was no risk of a fire getting out of hand. Eventually they settled down and the rushes got to be less frequent. As they had long days walking, they were tired by the end of the day and ready to settle down. If they were hungry or thirsty they had more of a tendency to rush, as they could not settle too well on an empty belly. They ate mulga and any grass as they walked along, but the water could be a day or so apart on occasion, as they were dependent on whether there were any bores, rivers or dams on the stock route. Station owners were never too happy to let drovers water their stock as it could possibly be detrimental to their own in the dry season, but if the water was on the stock route, the station owners could not claim it as their own.

Fred Crowther and his four men were disorganised and used to ask for food. They never seemed to carry much with them. One day one of the men rode into our camp and said to Mum, 'Can I have some tucker Missus?'

'All I got is a bone with a bit of meat on it,' Mum said.

'That will be fine Missus, we will eat the bone too,' he replied.

Mum gathered a few tins of food and suggested they make a stew out of it. Dad never begrudged a man a feed but if he drank, Dad was reluctant to help. This drover could not handle the wild cattle and they were soon handed over to another drover, Jack Pearce, whom Dad had known for a few years.

One of the stations we were crossing had a jackeroo showing us through and he told Dad about a boundary rider who had died on the station some time before. He was not missed for a few days as the land owners had been away shopping in Brisbane and seeing their children at boarding school. After they arrived home, they did not notice he was missing as he lived in a cottage through some trees that could not be viewed from the main house.

After the second night, the jackeroo had not reported in to the main house so his boss went looking for him. He was not at his cottage and his horse was not in the small paddock nearby. The boss went searching for him and found his horse at a gate to one of the outlying paddocks. After a brief search he was not found and the police were called in along with some neighbours to help search for him. Eventually, he was found hanging by his arm from a dead tree; his arm stuck in a hole. He was dead and had probably died from thirst. There was a parrot's nest in the hole and they believe he was robbing the nest for the young parrots. To reach into the hole where the nest was, he would have stood on the back of his horse and the horse could have walked off or possibly got a fright and shied, leaving the lad with his hand caught and no way of heaving his weight to pull his hand out. Around the tree were marks where he had tried to use his boots to get a grip on the side of the dead, straight tree, but there was nothing to grab onto with his slippery leather boots. His shoulder was broken and he would have been in agony each time he moved.

Robbing birds' nests was a common practice for station hands as feeding the chick and teaching it to talk gave them a hobby. Life on a station in the outback was not for the weak. It was a lonely life especially for those who lived on their own. Neighbouring stations could be quite a few miles away and in those days most station workers did not own a vehicle so they had to catch a lift to and from town with the owners.

One night I woke and went out to the fire to warm up. Our stockman Hughie was about to go on watch and he made a mug of Milo for both of us. He sent me back to bed after I had finished but having the Milo was not good for my bladder. I wet the bed. The next morning Mum discovered the wet bed and asked who had done it. Both Mary and I denied it, but Mum felt our knickers and I was found out. I got a hard smack on the legs for the lie. When Mum asked things like that, I figured it didn't matter if I lied or told the truth because most times, I still got a smack for it!

Hughie was not happy with the droving trip and he was not doing his job to Dad's satisfaction, so Dad gave him the sack. He drove into Dirranbandi and dropped Hughie off and he had to find his own way back to St. George. Dad walked into the café and there was a young part Aboriginal man who had worked with us before, Joe-boy Lamb. Dad offered Joe-boy the job and he accepted so Dad sent him out to his home, the local Aboriginal mission, to pack up his gear and be ready to leave when Dad called in for him. Dad picked up some required groceries including the meat and twelve loaves of bread then drove out to the mission. In those days, the missions were a mile or so out of town, a water hole with only the basic of comfort. People there lived in humpies and had no toilets, electricity or running water. It was not until early to mid-1960 that they were encouraged to move into towns or into a section of town with housing made especially for the Aboriginals to settle into. A pack of mangy, hungry looking dogs came rushing out to greet Dad when he drove in to pick up Joe-boy. Dad had bought us a bag of boiled lollies but he decided on the spot to hand them to the shy, skinny Aboriginal kids as he reckoned their needs were greater than ours. We kids were thrilled to see Joe-boy again. He was like an old lost friend and we much preferred him to the lollies.

The men were getting meat hungry and were sick of the tinned meat and fish that had become our staple diet. If we went through a village or Dad had to go to a town, he would bring home meat but it did not happen every week so it was pretty hit and miss. It was very difficult to get killers in cattle country unless you wanted to bring down a beast and Dad was reluctant to do this because we could only take the hind legs and a few ribs and the rest was wasted. Eventually, Dad shot a wild goat which was slaughtered and cut up. We sat down for dinner to a taste that was like nothing we had had before, it was very gamey. We ate some of that old smelly billy goat but Mum refused and so did Emmie. They both said the smell and taste was far too strong for them. Joe-boy showed us how to grill the goat meat over the open fire and that made it tastier, but Mum and Emmie still would not have any. The dogs had no qualms and had a feast.

Joe-boy as a lovely lad and we kids adored him. He was never alone when in camp as he had his own mob of inbuilt camp followers. If the men ever found any emu eggs Joe-boy would pierce a hole at each end of the egg and blow out the yolk and white. He would scrape images of animals on the shell, mostly kangaroos and emus. Sadly, living out of a truck did not fare well for this kind of delicate object to be kept for any length of time. Mum would utilise the eggs for cakes and omelettes for the men's breakfast. Emu eggs are very rich and taste quite rubbery. The cakes would be quite heavy but you would not hear us complain – even a badly cooked cake was better than no cake. Another of Joe-boy's skills was whittling, he was a real artist. He would get a gumnut and make it into a pipe for us. Of course, doesn't every boy and girl needed a pipe? He would get us to gather gumnuts for him to whittle into miniature horses, dogs, emus and kangaroos. I thought he was wasted as a drover's offsider.

Dad had to shoe the horses as the land was becoming quite rough with rocks and stones. He put the horseshoe into the fire until it glowed red hot and it was then plopped onto the horse's foot. The stink was awful but we all enjoyed watching this as the job was done in the camp. Dad had all the shoeing gear he needed plus a medium sized anvil he used to hammer the horseshoe into the shape of the horse's hoof. He carried a bag of various sized horseshoes in the truck along with the horseshoe nails.

As we only started out with one horse per man, Dad had had to buy extra horses off the station we picked the cattle up from to get the men out of trouble. Initially, they needed to do a lot of cantering and galloping to keep the stock in order until over time, the mob quietened down.

One of the stockmen, Alf Adare was a little, short, middle aged man with a great sense of humour. Beryl, being a toddler was at her cutest and Alf adored her. Each time Beryl toddled into sight after a sleep, he would say, 'Here she comes, looker ra.' Alf quite often rode a little grey mare called Iona and she used to buck him off regularly. He would go up in the air and land like a frog, on all fours. He had a weird way of horse riding; he sat straight up, his legs in the long stirrups, his hands sticking out the sides towards the front of the horse and he bounced. He would regularly crack the whip under the mare's belly and she would throw him each time. The men used to think it was a great joke and encouraged him to do it even more. Once he landed face first on a jagged log and he ripped part of his cheek deeply. Mum suggested he go to a doctor and get it stitched.

'Sew the bloody thing up for me please Missus,' he asked.

Mum cleaned the cut the best she could with Dettol and water, sterilised the needle with a lit match and, "sewed the bloody thing up". It healed with no problems but he was left with a two inch scar on the side of his face. He boasted to everyone how he got the cut and how the Missus sewed it up "good as new".

Early one day Col had just brought the horses into the camp. Beryl was still a baby in a pram and as Alf put the saddle on his horse, it pulled backwards. Alf got annoyed and yanked the reins. With that the horse reared up, Alf pulled on the reins and the horse lost its balance. The horse stumbled and knocked the pram over. It fell on its side but Beryl didn't cry; her big blue eyes were wide and round and she looked a bit horrified as if to say: 'Who did that to me?' Mum ran to the toppled pram and picked up the shocked, quiet baby out of the dirt and after reassuring herself that Beryl was okay, she rounded on Alf and soundly abused him. I was six years old at the time and I can still remember this short, chubby woman standing there, dress hanging below the knees with baby on hip, cursing Alf while the horse was still pig rooting around the campsite and over the fire; pots, pans and billy cans getting scattered in all directions. Alf was trying to hold onto the horse by the bridle while Mum was giving him a tongue lashing. I cannot remember what was the thickest: the dust, flying ashes and coals, or the language directed at Alf and his uncontrollable horse!

After this incident, all horses were banned from the immediate camp area. I guess the baby not crying was a fair indication of what was normal in the camp. With three brothers and three sisters doting on Beryl and wanting to push her in the pram, this was probably not the first time she had been tipped over.

Jack Pearce had his mob a day behind ours and one night Jack came to our camp for a visit and they sat up horse dealing plus anything else they could deal away. Jack always said he was going to wait for Emmie to grow up and he was going to marry her. Jack was referred to as old Jack so I am assuming he was well into his forties at the time as Dad was still only twenty-eight. In this dealing party there were two other men. They were swapping saddles, bridles, dogs, horses and each of the men ended up with their own gear in the end. Dad reckoned if they had a decent horse he would have swapped it for Mum!

After a while, Dad drove the truck into a nearby cattle station to ask if he could buy a killer from them but the station owner refused. Dad was annoyed at this as stations always had a few on hand to kill for their own use – beef and lamb. Most owners and managers did not begrudge selling the odd killer to a drover, as it saved the drovers from stealing and killing any of their "good" stock. Dad decided to "borrow" a fair-sized calf out of some stock that we were passing and he rode into the camp and got his rifle from under the seat of the truck, loaded it and meandered back to the stock that he had passed. He managed to ride right up beside the calf and shot it in the head out in the bush away from the camp. He then cantered back to the truck and retrieved some bags to carry the meat back to camp. The meat was hung off tree branches away from the camp so no passing visitors would see it. All the dogs had a good feed that night, like the rest of us. As you always let the meat "settle" after killing, the men decided that they would have BBQ ribs for dinner, so they lit a fair-sized fire and threw the cut up ribs onto the open blaze and coals and let it sizzle. Mum and Emmie did not enjoy the burnt taste so Dad grabbed a shovel out of the truck and grilled the ribs for them on it.

We kids thought that was great. Up until then none of us had ever had meat cooked like this. The men chopped the ribs into fair lengths of about a foot long. We all thoroughly enjoyed our "Fred Flintstone and his dinosaur bones" moment in time. I hate to think what we looked like after our tea as it was all very hands on with no knives and forks or hand towels. Our hand towel of choice was to wipe hands on our tops and skirts or in the case of the boys, their shorts. I can still remember the fun and camaraderie and laughter of that night.

Early the next morning, the meat was cut and salted. This was done by cutting long deep cuts into the meat and packing it with coarse salt, putting it in a sugar bag and hanging it on the side of the truck. The meat juice dripped out over a period of time and the meat became very dry but after it was boiled, it tasted much like corned beef. Fresh meat was put away for the next few days. Being winter and as the meat was very fresh, it kept for about three days wrapped in sheets and then put in the bed with the blankets heaped on top. Some of the meat was kept for the dogs and the remains were buried so as not to be found by any roaming, suspicious station hands – in particular the owner! A couple of days later Jack Pearce cantered up to us and told Dad the station manager who had refused him the killer had had his hay shed burnt down. Years later Dad admitted, he had galloped back late at night and lit the fire in the shed as payback for the petty refusal of the killer.

We were in red, sandy country now and we kids were playing while waiting for the stock to catch up with us at dinner time. We came across what we thought was a lizard with a fat tummy. We poked at it to try and make it move because normally a lizard would run like crazy from us. We very rarely got close to any kind of living creatures as we were far too noisy for any self-respecting lizard or snake to stay put for long. Mum heard us carrying on over this fat "lizard" and came to inspect it. She received a shock to realise we were poking sticks at a Death Adder! This was quickly done away with by Mum wielding a shovel. Les cut open the short, ugly monster and it had a full bird inside. We found this fascinating as I guess we never thought about what snakes ate apart from mice. None of us had picked up on the fact that the "lizard" had no legs!

Eventually we got closer to Charleville and the "jump up" became a conversation piece. The jump up was actually "land earth movement" that is quite prominent in various parts of Queensland. The Angellala jump up was only about 100 metres high at the most. (After a web search and local Charleville knowledge, I found out the jump up was called Etona, the area of Angellala – there is also a river called Angellala that we would have crossed, hence our confusion).

We were warned that the truck and caravan would not be able to be driven up it, as the road was badly eroded in places and no vehicles had been on the track for several months. The road we were on was a bit of a goat track itself and Dad suggested to Mum she should drive the truck and caravan around some miles to Morven and come back via Sommariva and meet them with the stock at the top of the hill.

Mum's sense of direction was worse than Dad's and she flatly refused to do it. Dad could not afford the time to make the long trip himself so he decided he would take the truck and caravan up the jump up after all. It would have been a three or four hour long round trip to come back to the road at the top of the hill. Les rode with the stock as an extra man in Dad's place. Dad then rode up the steep jump up to see if the vehicle could make the trip. He thought the truck and caravan would if he did some minor road works. We were busy doing the road works when Dad thought a shovel, wood saw and crow bar was needed, so he sent both Mary and me back down to the truck to get them. Instead of walking on the gravely steep road, we ran flat out and in seconds of each other we fell flat on our faces. We took skin off our knees and elbows, it stung badly and bled a bit. Upon arriving at the van, Mary fished out the can of goanna salve and put it on our grazes. We had a beaker of water, retrieved the tools that were needed and sedately walked back up the hill.

Working together, we slowly made our way up the steep grade. Gathering rocks, sticks, logs and stones, we filled the ruts and washouts to help get the truck and van up and over the hill. After a brief period of time Beryl got to be a bit of a nuisance, toddling off to all places. She fell down and rolled a bit over the edge, screaming blue murder with fright, Mum accused us all of not looking after her properly but we all had our bums up and heads down gathering sticks and stones to help fill in the gullies. Mum decided to send Mary and Beryl back to the truck and caravan so we could get on with our work. Beryl was not the only one to roll over the edge. As the best rocks and large stones were on the edge of the graded road, we had to get close to the edge or stand on the edge to pick them up. Of course we slipped a bit, but as there were trees and shrubs close by, we did not slip too far. If we did, we would crawl back on all fours and get on with the job. At an early age we were all taught not to be sissies!

After the bridging work was completed we walked back down to the truck and caravan and had a drink and smoko. We were filthy with dust and dirt. Dad sawed up a log and gave Mum a decent sized piece to carry and Emmie carried a smaller one on the opposite side of the truck in case they had to use them to chock the wheels of the truck if it lost traction and started to roll backwards. Then Mary sat in the passenger side of the truck cabin to keep an eye out on that side and Mike and Beryl were in the cabin with them. The reins of the grey mare were handed to me to lead up behind the truck and caravan and away we went. I walked quite a few yards behind the caravan leading Blue Bell, trusting the truck not to slide backwards on me and the mare. In places there were a few scares where the truck lost grip and started to slide. Mum and Emmie would quickly put their posts behind the back tyres until the truck gripped and Dad carefully and slowly took the truck up the steep incline and on occasion the truck would spurt out some dust and dirt, but it all went smoothly enough and when we got to the top, we had a well-deserved rest. Dad rode back down the hill to help the men bring the stock up. Being cattle, they walked on the track and beside the track, wherever they wanted to really. After we did the road works, word got around that we had taken our vehicles up and over it and then other people along the track started to use it again. If a drover could take his plant and stock up the steep incline, so could they!

We finally went through a cattle station called Lesdale, near Charleville and this was our last stop before the saleyards. The next day Dad and the rest of the men had to put the cattle into the cattle trucks to be trucked up north. It was Col's job to get into, onto or crawl along the sides of the cattle trucks to hunt the beast up to the front so more animals could fit in. He had a long stick that he could either poke the cattle with or give them a whack to make them tighten up to the others. It was hot, dirty, dusty work and very noisy with dogs barking, cattle bellowing, men yelling and whips cracking. As the trucks were filled, the gate then had to be closed. The full truck was driven off and another empty truck backed up to the race and loaded. The whole time the cattle were being loaded onto the trucks, the air was thick with red dust, a pall of it lingering all over the town. Any slight breeze was gratefully accepted by all.

Charleville saleyard was on a flat red sandy area and about a mile out of town. The local Aboriginal kids came to visit and take a look at the family of white kids and the array of dogs, horses and other gear that we had. We enjoyed playing with the town kids if they biked out to visit. We did not often get the chance to do that as Mum was suspicious of them stealing – we had nothing of value that they could steal anyway but that was camp law, made by the camp boss!

BURREN JUNCTION 1958

Traditional land of the Kamilaroi People

Dad was offered a job to graze 1,500 head of sheep along the stock route while the station owner mustered the rest of the stock to be shorn. We picked up the stock from a small place called Come By Chance, south of Walgett. Dad was to drove them slowly for about three weeks and bring them back to the home station, *Nevertire*.

As the stock route was a large lane, Dad decided that he, Col and the dogs could handle them without extra help. We arrived at a small town called Burren Junction with the stock, the local policeman drove out to give Mum the news that her mother had died in Sydney. Dad put Mum and Beryl on the first train out and he was left minding the six kids: Col aged eleven, Emmie ten, Mary nine, Les eight, I was seven and Mike five.

For ease of handling, Dad let the stock go in the open lane. He had the idea that they could be left there with dogs tied each end to keep the sheep in. Later in the afternoon, he decided he had better put them in a sheep break after all. He took Col and Emmie to help and of course, the truck and dogs went too. A deep hole was dug in the hard black soil and a small fire was lit in it. The grass was cleared for about two feet around the hole to reduce the risk of the fire getting out of control. The Mary, Les, Mike and I sat around the piddly fire with nothing to do. All we had with us was Col's camp bed and some camp gear.

All went well until it got dark, which happens very quickly out bush. We started to tell ghost stories, the scarier the better. It soon became very dark and we all became very, very scared. We could see the truck lights in the distance and hear the sound of the hammer banging in the pegs for the sheep break so we decided to run towards the sound of the motor and the lights for the extra security of the family. For protection, Les picked up the axe, Mary the shovel, I had a big stick and Mike grabbed a twig in his little fat hand. We ran. Mike lagged behind a bit so we encouraged him to run faster. Unfortunately, between us and the truck was a bore drain and Les splashed straight into it. We heard his yell of annoyance and surprise but we were not stopping for anything because we could feel the ghosts breathing down out necks! After finding Mike in the dark, Mary and I grabbed a hand each and carried him between us. We found the road and ran along the best we could with our "self-defence" gear at the ready. We could hear Les gathering ground behind us, so we knew he had not cut his leg off or been taken by ghosts and eventually, he overtook us.

Dad, Col and Emmie were all surprised when we turned up yelling and screaming. Dad's biggest worry was the fire could have spread but we were too "bushy" to let that happen. We had put the fire out before we left the camp by pushing soil onto it. I had learnt at an early age water and wood were often in short supply and knew it had to be used sparingly. But it was now pitch black and with no fire, Dad had to look for the wheel tracks going off the side of the road to try to find the campsite. He was really angry at us. I was petrified, waiting for the smacks that never came.

The steam trains stopped to fill their tanks with water just outside Burren Junction and this was fascinating to watch. The men worked like they had no fear, deftly climbing on top of the high water tank to drop down the hose to fill the steam engine to carry them along to the next point of call and another watering point. Sometimes the train drivers would blow the horn and wave to us and we would all wave to the drivers and guards and most of the people on the train would wave back. The train driver and passengers often threw a newspaper or novels to us.

While Mum was away Dad also had to cook for us and do the washing. We collected our dirty clothes and went to the bore head with him while Col stayed with the stock. The bore water was hot when Dad bucketed it up and he poured it into the tub on the fire then threw our dirty clothes in. After leaving the clothes to merrily boil for a while, he then bucketed water out. He had a look of horror on his face as he saw the water had turned red and the clothes had all been dyed various shades of red and pink. Dad didn't live that down for a long time. His shirts and underpants and the boys' clothes were pink! In those days, real men did not wear pink.

After several days a train came along and a woman with a young child hung out the window and waved and shouted. Mike and I were placed in the truck and off we went to Burren Junction to get Mum and Beryl off the train. But there was no Mum or Beryl to be found. After searching the railway station and the town, we dejectedly returned to the campsite, realising we had chased after strangers who had waved to us. I think Dad was quite lonely during this time with only six kids for company. A few days later another train came along and there was no doubting the shout of the woman and child belonged to us, off we went again. We were all pleased to have them back with us.

WEEMELAH - BANGATE STATION 1958

Traditional land of the Kamilaroi People

Dudley Olsen was an old friend of Dad's who won a Return Soldiers League allotment he called *Vauxhall*. The block of land was on the Brewarrina Road about three miles out of Walgett. Work was scarce again so we camped in the Olsen's front paddock, just inside the front gate. We were far enough away from the house to have some privacy and close enough to the river to collect water for the dogs when needed.

The workmen on the properties or stations were supplied with free housing, fresh meat, eggs and milk and if the station was big enough, free vegetables. The jackeroos were quite often fed at the main house though their accommodation would be a bit of a distance away for privacy. Their wages were very low though and if you saw a young man driving a car you automatically knew it was the cockie's (boss') son. Around Walgett and the surrounding towns the fences beside the grids on the main roads had tin cut-outs of various Australian animals hanging on them. Apparently, the jackeroos did these in their spare time to relieve the boredom. These cut-outs were quite large, around five foot (152 cm) long and about four foot (120 cm) high. There were emus, kangaroos, dingos, cattle and horses. Travellers driving past would stop and take photos.

Dudley and Helen Olsen had four children, Beau (Maurice), two adopted children Linda and Carol and then along came Cathy. Theirs was a rather unusual story. Beau was about eight when they decided to adopt a baby, but the only babies available at the time were sisters whose parents had been killed in a car accident. The Olsen's agreed to have them both. The girls came home and they became a happy family of five. A few months later Helen was sick in the middle of the night with bad stomach pains. Dudley bundled up his little family and drove them all into Walgett's hospital and not long after Helen gave birth to a beautiful red-headed daughter they called Cathy. What a delightful surprise for them all!

When Cathy was about two years old, she went missing. Helen searched inside then began frantically calling for her outside. The Castlereagh River ran beside their house and it was flowing strongly. Helen feared the worst and was about to call the men to come and help when she saw Cathy down the road walking towards her. She was holding onto the family dog with one little arm wrapped over his neck. Upon rushing to the child and dog, Helen became aware that both of them were wringing wet. She never did find out if Cathy had fallen into the river and was rescued by the dog or if the dog had been rescued by Cathy!

Beau was the same age as me but he was very mature for his age. Beau, Col and Les would jump on the back of the Olsen's ute and go kangaroo shooting with Dudley driving. This was quite acceptable as the kangaroos had to be kept to a manageable level because they competed with stock for grazing.

At this point, and to make ends meet, Dad was still collecting copper, brass and used batteries for resale. He had also gathered horses, mostly old useless ones not worth much in that he would send them off to Parramatta, Sydney for sale. We called them "doggers" and we believed they were used for dog meat. We found out much later they were also used for human consumption. The sellers of the doggers had to get a railway truck full to make it worthwhile, this being 200 horses. Dad had a friend near Weemelah some eighty miles from Walgett close to the Queensland border, who owed him a favour and he gave Dad use of a paddock that was full of washaways. This is where Dad kept the doggers until he had enough to sell. Dad drove around far and wide, including into Queensland, buying or begging for horses that he could sell to the knackery.

One day Dad dropped a truckload of horses into the horse yards that he and Col had made and he decided to muster the rest of the paddock into the yards. One of the horses had broken his front leg. Dad had to shoot the horse and sadly, he did it in front of us kids. Hearing the loud bang of the shot, I turned and witnessed the horse falling to the ground, kicking his legs wildly. I found this very confronting and cried for days afterwards and have always remembered that scene.

This job of collecting horses and selling them was necessary until Dad was offered a droving job, which was his first love. Even when he was on a droving trip he continued collecting horses and kept them in his herd of stock horses until he had a chance to truck them to Weemelah or if there were enough he would get Col to drive them to Weemelah and we would follow in the truck. This took a few days depending on where the droving trip finished. If we did not have another job to go to, it was quite viable to do this, costing only the fuel in the truck and the time – of which we had an abundance.

Many good horses were found amongst these "yangs" (horses that were worthless to most people). If Dad liked the look of a horse, he would keep it aside to retrain. He probably made more money wheeling and dealing with the better horses than the actual sale of the doggers.

Dad found a pony he wanted to break in for Emmie to ride. He spent a bit of time trying to quieten the pony down and after a few days he tied it to a tree on a long rope. He then ran at the pony, waving an old oat bag and yelling. The pony got such a fright it tried to gallop off and the rope tangled around its front leg. It toppled over and broke a leg. Dad walked to the front of the truck, retrieved his rifle, put a bullet in it and coolly shot the pony between the eyes! A rather sad and traumatising thing for us to witness, as once again it was done in front of the family. All horse breaking and training was done near the camp so at an early age we had to toughen up to the facts of life.

I was about seven when disaster struck… the NSW Education Department got in touch with my parents and told them if we were not sent to school, they would be charged for not educating us. This was what happened when drovers stayed in the same place for too long, so a decision had to be made. The first thought was to buy a house in town so we could be sent to school but Mum refused. She wasn't going to live on her own in town. They decided we would do "Correspondence" with Mum being the teacher. Our names and dates of birth were sent off to the officials and samples of school work were sent to us so we could be graded.

What a nightmare! None of us could read, write or spell so we were placed in grades in twos: Mike and I, Grade One; Les and Mary, Grade Two: and Col and Emmie were Grade Three. Emmie took to schooling like she had been doing it all her life. She learnt a lot while sitting in the front of the truck with Mum and Beryl, Mum being her constant tutor. Col showed no interest at all and Les and Mary's attitudes were much like Col's. Mike and I, being of such tender years, were bullied into it and when the school work arrived I was excited to start learning, although the repeated Dick and Dora routine got to me. I found I had an aptitude for schooling except for maths, which I hated with a vengeance. So began the monotonous rote of times tables. If at any time we were to be punished we had to do the times tables. I didn't ever learn them very well. To be honest, Mum was not much of a teacher, and her ready use of a wooden ruler did not help!

It was not long before Col did not do any school work at all. Emmie did his with a few of the obligatory errors. As we moved so often, we did not get much time to sit and learn so it became *ad hoc.* After getting in trouble with the Education Department for the school work being sent in late, Mum would sit down and do our work and we would copy it with a few errors to make it look authentic. We did this to the end of our school days.

Because we drove stock between New South Wales and Queensland, we had to change to each state's style of schooling. Both had completely different hand writing, which I found very confusing. The correspondence teacher suggested we each write to a pen pal to make writing more interesting for us. Mum and Emmie helped us compile our letters and doing the compositions regularly helped, although it was still a struggle. Emmie wrote to Col's pen pals for him and we others struggled along, although it did not take Les long to drop his. After I started to learn to read with a bit of competency, I began enjoying the compositions, although Mum wielding the ruler with cruel regularity did not help me much. We were afraid to ask her questions as each word she spoke was followed by a hit with the ruler wherever she could reach – head, arm or leg.

Emmie made the mistake one day of asking, 'Mum, why don't we drive ahead to the campsite and do our schooling instead of driving slowly behind the sheep all day?'

Mum smacked her over the head. 'Mind your own bloody business,' she said.

Emmie didn't mention that idea again! It was the first and last time I saw her get a smack.

Dad was offered a job to take a mob of 4,000 freshly shorn sheep from *Bangate Station* between Goodooga and Lightning Ridge to a station near Tullamore, a small village out of Peak Hill. Dad once again hired Alan Alston who was in Walgett and his friend Fred Price to do the trip with us. As we were still in Weemelah at the time, he collected the horses he required for the job and he also chose five extra horses he was training to be stock horses for future sale. Mum put Col's lunch into his saddle bag and gave him a water bag to take with him as he had to drive the horses to Goodooga, some miles away cross country. Dad gave Col his watch and gave him a crash course in telling the time. He told him to the take the horses in a trot or light canter to save their energy and to change horses on the hour until we caught up with him. He was also given instructions on which roads to take so we would meet up with him along the track, as we had to go via Mungindi to pick up the two working men. Poor Col was only about twelve at this time but he cantered off confidently, holding his horses' reins in one hand and a long-handled whip in the other. The horses went into an easy comfortable canter, with the dust flying behind them.

Mum watched him for a while with a worried frown on her face, but she had to put her concerns aside because we still had to finish packing up all the camp gear and collecting any rubbish lying around. Then we drove into Mungindi to collect the men from the local pub, which was a meeting place any country man could find. Of course, Dad had to have a drink so while they slaked their thirst and caught up with the gossip, Mum did the grocery shopping for the trip. After she finished this chore, we played the waiting game until the men's thirst was satisfied. After an hour or so they came out, Dad stepped into the truck and off we drove after Col. Mum gave Dad an earful as he was nearly drunk and she was worried about Col being on his own for so long. After considerable time driving cross country and black soil roads that were only a two-wheeled track, Dad finally picked up horse tracks. Col had managed to follow the instructions fully.

Col and Fred Price shared driving the horses to *Bangate Station* over nearly two days in easy stages for the horses. They willingly galloped along the side of the road, kicking up their heels and having a little buck on occasion, full of life, their tails swishing and their manes flowing in the wind. Dad had at some stage bought a white albino horse that had a red head and neck. He was also in this mob, making a show of fine horse flesh. We called him Redhead. He could not see very well and would trip over the tiniest thing but we kept him for many years. He was one of those "emergency" horses, like the useless dogs for "just in case"! We drove ahead of the horses and put the billy on so the horses and riders could stop and have a rest. Then a fresh horse was caught and another rider would saddle up and travel to the next stop. The two days passed by pleasantly, with us kids lazing in the back of the truck doing very little.

Bangate is an old sheep station that was one of the original homesteads from the early 1800s on the bank of the Bokhara River, which runs into the Darling. The riverboats used to come right up past there moving people, wool and groceries that were needed for the duration of time until the next riverboat. If the river was not running, which in dry times it did not, the people on the station had to ration their food to last from six months to a year. Stations of this size would have a collection of local Aboriginals for house staff and stockmen. They would provide their clothes, which were brought on the riverboats and only buy the staples such as sugar, flour, salt, syrup and treacle. A vegetable garden supplied their fresh food needs. It was horse and cart only to the nearest town, Moree, which was a three-day ride away.

More towns were being settled as the wide open spaces became more populated. Some enterprising people would open a store and either have their own wagon to bring in the goods or they would rely on the bullockies to bring their requirements when they dropped the loads of wool off at the nearest railway station. Moree was closest until the line opened up in Bourke. The bullockies could not work in the wet, as the wagons were too heavy to be pulled through the mud. When they were bogged, the bullockies stayed with the stock and lived off the land after their provisions ran out. They would catch and kill wild turkey, kangaroo, goanna, birds and anything they thought was edible. This was a life-threatening event.

Dad had decided to take the sheep from Goodooga to Lightning Ridge, Walgett, Coonamble, across to Nyngan and on to Tullamore. It was a long trip and he only knew half the country that he was going through. Many droving jobs took men to places they had never been before.

In the summer, Dad liked the stock to be on the road not long after sunup and into the night camp just on sundown. The stockmen's day did not finish until the sheep were in their break. An early start gave the stock time to graze as they walked along and this included a spell of some sort during the day if they were so inclined. Summer in the Antipodes is December, January and February and it is very hard to keep stock moving in the extreme heat as most days it was up and over 100°F (38°C).

If grass was not in abundance, the riding horses for that day had a few handfuls of chaff and oats in a nose bag. A nose bag was a sugar bag with one end open and on each side was tied a piece of rope long enough to go over the horse's head and ears. It hung half way up their face. To reach the grain, the horse put his head down, the bag touched the ground and the horse could then use his lips to gather up the grain. It was not uncommon to see the horse shaking his head to get the odd bits of grain to fall off the side of the bag if there was any there. They were given a drink out of a wheel cap and the horses would have their lunch while their riders had theirs. Even when the sheep were camped, the men still had to be around them as some could meander off, looking for juicy grass to nibble on.

The stockmen would lie under a tree or in the horse's shadow and have a rest, placing their hats over their eyes and maybe even having forty winks. They'd rely on their horse or dog to let them know if the sheep were on the move. Sheep are inclined not to walk in the heat, so the idea was to get them to cover the biggest distance in the morning, hence the early start. They would rest during the hottest part of the day and as it cooled off a few degrees in the afternoon, they would be inclined to walk along and graze, hoping for a drink of water before being settled down for the night. A droving job is a seven days a week, twenty-four hours a day job and a good drover was up and starting their day before daybreak, with breakfast being eaten at the break of day.

On the really hot days the dogs would have to work extra hard to keep the stock moving as the sheep would only want to walk from tree to tree, stopping in the shade. We kids would also be put to work to help move the sheep along. We used a tree branch with heaps of leaves to push at the woolly beasts to try and keep them walking a few paces and we would also use tin rattles made from small milo or milk cans. Inside the cans we would place small stones or our precious marbles to make a noise and we would rush at the sheep rattling the tin to give them a bit of hurry up. This was hot dusty work, but we mostly enjoyed it as we felt we were a part of the working team helping the men. We slept well at night after exercise like this!

We were so poor, we kids only had one hat between us, which had been handed down from Dad. If we had to do stock work, whoever we were replacing would give us their hat to wear. I gather that the sun did not do any damage unless you were on a horse or walking behind stock! If the men replaced their old hats, these were handed down for us to wear. Being bushies, the hats were worn out before being handed down but they did give us shade. Most of the time we looked like legs with a hat on as we were so small and the hats were so big. We always managed to lose our hats. We would put them down and never remember where and mostly they were not missed until the next camp. By then it was too far to go back and look for an old worn out hat.

Dad was a red head and light skinned. He never went anywhere without a hat. As soon as he went out of the truck or the caravan, he would place it on his head. Some nights he would walk out of the caravan and automatically put his hat on and if he was sprung by one of us, we had a good laugh and would ask if he afraid of the moonlight.

The moonlit nights were nearly as bright as days, so if it was very hot, Dad brought the sheep in later in the evening in the cool. Settling the sheep down into the sheep break at night was quite manageable, but this was not encouraged as the men also needed their rest after working long, hot weary days. Daylight started at 4 am and it was not dark until 7 pm or later. On these long hot days, the camp would be bedded down straight after the night meal was finished and the dishes were done.

When it was so hot, the dogs were taken out of their crates and tied up under a shady tree or along the side of the truck, so they had some shade. They had to be tied so they could not reach each other because some had a tendency to fight. They were always given a drink of water out of the trusty wheel caps and a handful of dog biscuits. Before we went to bed, I would carry a bucket of water over to give them another drink. If needed, Beryl would carry a lantern and Mike would carry three hub caps so three dogs could be watered at the same time. This made the job a bit quicker.

Straight after breakfast, Mary and I would wash the dishes and put them away, then along with Mike we would rush to put the dogs into their crates. If one was a fighter, a muzzle was put on him first and then possibly he also had to be tied to the side of the crate or he would deep trying to fight the others. The chains were removed off the dogs as they went into the crate and it was not unusual for a fight to start as we were driving along. They all barked like mad when the truck was moving. If it sounded serious, Mum had to pull up to the side of the road and we would try and stop the ruckus. This entailed opening the crate door and trying to pull them apart and of course we quite often would get bitten and they would have a lot of blood on them too. We only lost two dogs in this way while driving along and that was when we were on a long trip going to pick up a mob of sheep and we did not hear them fighting, possibly because we were all talking. There was always a main offender so he would get the hiding and be muzzled and then be put back in the crate.

During summer we had to be extra careful with water. The men would take their hats off and push the crown in to make a bowl and water their dog. Occasionally their horse would also have a mouthful – a nudge from the horse was the hint! We only had three 44 gallon drums of water to do the whole plant. The third drum was tied up under the back of the truck and had a tap on it to make getting the water out easier. We had to siphon the water out of the other two drums in the back of the truck. We had a length of hose about five-feet long and that was put in the bung hole. One of us would then suck on it to siphon out the water into a bucket. We were always careful not to spill any. Dad wound some wire tightly around the end of the hose with a loop on it so it could be placed onto the side of the truck, high enough so the water did not run out. It was a tragedy if the hose slipped off and emptied onto the ground without anyone seeing it.

If possible, the horses were watered every afternoon, even if this meant taking them to a nearby station to do so. The horses and dogs were the first priority in any drover's plant. Most droving camps were by water on the stock routes when not on private land. The public dams, windmills and bores were placed 10 miles apart as the early drovers mainly moved cattle and the water was placed for them. The sheep had to travel for more than a day to get to a water supply. The government learnt as the country was opening up, water had to be supplied for the drover to water their stock. Putting in the water supply was a great cost but if they had not done this, rural towns would not have grown as quickly as the government of the day wanted them to. This also created more jobs in the rural sector as after a bore was drilled, the bore drains had to be established and these ran for miles over neighbouring stations.

Alan Alston was a great worker and friend to us all. He was very patient and happy to show us how to do things that Dad either did not know how to do or was not bothered to teach us. Alan was good at whittling and would make us pipes for pretend smokes or scrape the paint off the top of Log Cabin tobacco tins, making pictures for us. On these tobacco tins, there was a cabin and a man sitting on a fallen log holding the reins of a horse. Dad would play a game with us and these tins. He would ask, 'Where is his dog?'

We didn't know as we could not see any dog.

'There's his dog,' he would say, 'behind the tree having a shit.'

This caused great gales of laughter as we were not allowed to swear. We were camped close to Walgett one night and Alan, who was quite fond of a local waitress, wanted to go into town for a visit. Dad agreed for him to take one of the horses to go do his courting. Mum needed some groceries and as Dad needed fuel, she sent Mike and me to town with Dad to do the shopping. She obviously needed a break and we did not mind as it meant a trip to town. If we were lucky, we knew we might even get a small bag of lollies to share or an ice cream. One of Dad's party tricks was to try and spoil Alan's love life and Dad spied him in the café talking to his sweetheart.

'Patsy,' Dad said. 'Walk up to Alan, grab hold of his hand and say, "Can you buy me an ice cream please Daddy?"'

I did this and Alan went several shades of red, spluttered and looked from his sweetheart to me and back again.

Dad came in laughing away at his joke. He did this sort of thing regularly to Alan, using whichever kid was available and at different towns and cafes. Alan was a good sport and took it all in fun and we kids enjoyed being part of the fun.

I recall a time when Dad spied some fat sheep grazing in a paddock and he fancied a few as killers, so he sent Alan after them. Showing off, he galloped towards the sheep, the horse tripped in a rabbit hole and fell over, causing him to fall off his horse. Somehow when it was getting up, the horse trod on the side of Alan's head and tore one of his ears half off – it was just hanging there. Alan rode up to the truck with blood pouring, holding a dirty handkerchief to the side of his face. Dad and Mum wanted to take him to the nearest hospital for treatment but he would not hear of it. Mum cleaned up his ear the best she could and put a few stitches in it herself. With his head wrapped up he looked like a pirate, much to our delight. Mum made him take a couple of Veganin headache tablets and put him to bed in the back of the truck for the rest of the day. The next day he was back on the horse, doing his job without complaint. Living on the road with the dust and dirt and sweat and no regular bath, it was a miracle that it did not get infected. Thankfully, his ear healed well. They don't make men like that anymore! As we met him over the years he would proudly point out his scar and he always thanked Mum lavishly for what a great job she did. Another one of her medical success stories!

One night Mum got some lamb chops out to cook for tea but she discovered they were fly blown. So she went to throw them to the dogs but Alan stopped her and he bet Emmie ten shillings he would eat them and the bet was taken on. He closed his eyes and with pretended gusto ate two of them. Dad paid Emmie's bet for her, he said it was money well spent. It was not uncommon to get out the salted meat to cook and find maggots in it. We would just wash them off and cook the meat anyway. No workplace health and safety back then – or finicky kids and workers.

There was a woman drover called "Old Wild Rose" in Dirranbandi, who had a wagon and used to go droving with her sons and another bloke called Len Frite from Walgett. Old Rose used to have a shot gun and shot one bloke in the backside for giving her cheek one day. This Len Frite was a big guy and a real standover merchant and skite. He met his match one night in Collarenebri, when he went into the local pub and mauled the barmaid by hugging her and patting her backside, skiting in a loud voice trying to attract attention. He walked over to another chap called Payne and took his beer. Payne said, 'Hey, that's my beer, leave it alone.'

Frite replied, 'Don't you realise who I am? I'm wild Frite from Walgett.'

Payne replied, 'Yes and it's gunna be a quiet night in Collarenebri.' He then proceeded to knock Frite down with three quick punches. Frite got his reputation from hitting smaller men and Payne was smaller than him but very solid. Frite would hit people with a stirrup iron, whip handle or anything he could get his hands on. After that incident someone wrote a ditty about him: 'Len Frite, Walgett skite, never could ride, never could fight.'

Dad was in a pub in Walgett and a bloke who used to hang around with Len Frite took his beer and drank it. The next shout Dad took the bloke's beer and threw it in his face. Dad was nearly beaten to a pulp for that. Being a drover in those days meant work was found at the local pubs, so sadly Dad had to spend time at the pubs. He never backed down from a blue and if one was on, he gave as well as he got. He was short and wiry and at his word, never lost a fight, except for that one in Walgett! In the mid-fifties Walgett was a rough place to be a stranger. The local drovers wanted the work for themselves and did their best to try and keep the non-locals out of their territory. They used scare tactics, sneaking into camps at night to slash their gear, kill the dogs, and shoot their horses. One drover got his plant burnt out, so their beds, harnessing, food etc. were all gone. This particular drover only used saddle packs and they were burnt too. Luckily Dad had been warned before going there so he was prepared for the rough treatment. If you looked like you were going to fold, they got worse but if you stood up to the bullies, they were not too bad.

Dad swapped a horse for a smart little white pony for Les. Les named him Captain and adored him. At around the same time Dad bought Les a small dog called Snip and the three of them went everywhere together. Les was our parents' pet and consequently, he was spoilt rotten. He rarely did anything he was asked to do and neither parent could control him from an early age, so they just let him do whatever he chose to do. One particular day we were at camp and Mum told Les to give the rest of us kids a hand to pick up some wood for the fire. He flatly refused, Mum lost her temper and threatened, 'I'll tell your father when he gets into the camp.'

Les said , 'You help them, you old bitch,' and he ran down to the river bank.

Mum immediately lost her temper, leant down, picked up the axe near her and flung it at him. It went swinging in the air and the handle hit him just behind the knees, hard enough for him to trip over. I guess it was lucky in the timing as the outcome of this could have been a whole lot worse. Les stood up, looked at the axe, looked back at Mum and put his thumb to the side of his nose and waggled his fingers while poking out his tongue and chanting, 'Nah, nah, nahnah, nah.' He then kept running down the river bank out of sight. After the sheep were placed in the break for the night, Mum told Dad what Les had done. Dad pulled his belt out of his trousers and said to Les with a threat, 'Come here.'

Les grinned at him and stood still.

Dad took two steps towards him and Les took a running leap to Captain, jumped on his bare back, gathered the reins and galloped off into the sunset, laughing.

Dad shook his head at Mum and we got on with getting dinner ready. Just as dinner was nearly finished, Les rode back into camp. Mum handed him his dinner and we all sat watching closely, expecting Dad to pounce and give him the flogging of his life but disappointedly, he seemed to have forgotten about Les' bad behaviour.

One evening Les refused Mum's request to water the dogs and this infuriated her. When Dad came in later she was fed up with the lot of us. She told Dad the story and he took off his belt, made Les bend over and really gave it to him. Les let out a howl, jumped on his horse and galloped up the road and cried and howled and screamed and carried on. Eventually he rode back into camp, sniffling and feeling sorry for himself. In the meantime, we other kids had watered the dogs and done the usual chores that we had to do. This would have been the first and last hiding Les ever had and I think it says a lot about how tired Mum and Dad must have been at that time.

Dad was still on the lookout for any yangs for wheeling and dealing, and the station manager had a particular horse that the owner did not like. The horse was difficult to get on and off, would not canter, was hard on the bit, refused to lead and had other problems as well. Dad traded one of his horses for this one and the station manager and Dad were both happy with the deal. Dad asked the men and boys to ride the new horse till he stopped bucking and pig rooting. Dad taught him how to lead properly, clipped him all over, brushed his tail and clipped his mane and he proved to be a really good horse. A year later as we were going past the same property, Dad sold this same horse back to the manager – and for a good profit! Going through the same station sometime later, the owner rode him into the camp. He and Dad had a good laugh over his "new" horse that was so good, he had bought him twice!

We all wore our clothes for days on end and did not bathe very often, but when we did, we had to carry two buckets of water from the river or trough nearby. The first bucket went into the tub cold. The second one was heated on the fire before putting it in the tub in the back of the truck. A blanket was placed across the sliding door and we took it in turns to have a bath. The bathing started in the late afternoon with Mum first, then Beryl and Emmie as they were the cleanest and then the rest of us at the camp followed. The men came in one by one after us so they could have a bath before it got too late in the evening. The water was not thrown out, it was just topped up with a small amount of hot water for each person. By the time everyone had finished it was so thick, we could have thrown in some stock cubes and made a soup.

If we were not near water and it was time to bathe, we would siphon water out of one of the 44-gallon drums and with probably no more than 3 inches of water, we struggled bravely to get clean. Normally no one missed a bath. If we had workmen, they were pressured into having one also but they could be reluctant bathers. If we were near a river, dam or bore drain, the workmen preferred to have their own ablutions there.

On most stock routes you rarely saw other vehicles so it was quite a relaxing time. Out in a paddock, wild sheep would gallop away when they heard a vehicle coming but if there were vehicles, one of the riders would guide them through the mob. The horse walked in front of the car with the dog nipping a leg or two and the sheep would part to let the vehicle crawl through. Some people would rev their motor up, trying to hurry things along but most people were considerate. Impatience was a curse with the drivers but truck drivers were the worst. Some would blow their horns, shout and curse at the drovers and make a general nuisance of themselves.

Along the highways, the lanes were a lot wider so the stock could be kept off the road altogether. This was sometimes difficult as the sheep knew there was sweet grass growing on the side of the road and they kept trying to get back to it. Once the dogs knew where the sheep had to be, they would be alert and keep them back if they made a move towards the road. One particular trip, the sheep were spread out along the road, feeding as they went and a car came travelling really fast, not taking the stock into consideration. It hit four sheep, killing two and maiming one so badly, it had to be put down. Dad saw the incident and galloped up to the car, which had stopped. He jumped off his horse and grabbed a stirrup iron and leathers off the saddle and threatened to kill the driver, he was so wild. The driver paid the exorbitant price demanded of him for the sheep and apologised profusely. Dad did not calm down and told him to never show his face anywhere near him again.

Not long after this incident, another car was coming along so Col sent his bitch Cindy to get the stock off the road so the car could keep going. The sheep came rushing back across the road and Cindy hesitated on the opposite side looking at Col for further instructions.

Getting none, she raced back to Col, straight into the oncoming car and was killed instantly. We all missed her as she was a good loyal working dog. Over the years we lost two or three more dogs due to speeding vehicles.

After dinner one night, Dad drove the truck to a station to get the water drums filled and the lady of the house had a handful of kittens. She offered Dad one for us kids and he came home with this little bundle of fur. We all loved and spoilt this lovable, adorable creature. Beryl called it Cuddles. It was grey with a white blaze and four white socks. There were many fights over who would cuddle, hold and feed her. It grew up to be a beautiful cat, but one night when we had to go to town to get something, Cuddles got out the back of the truck and sadly could not be found. A sad time for us "little" ones as we adored Cuddles and smothered her with love. Who would love us now? I fretted for months over that kitten.

One time the stock were going through a large sheep station when two young Aboriginal men turned up on horses to escort the men and stock through, so our stock did not get mixed up with theirs. The two lads rode in front of our mob on the wings and if they saw any of the station sheep, they would send their dogs to hunt them further afield. At lunch time they rode to our dinner camp and Dad asked them to come and have a cup of tea with us. When they did, us kids stared at them as we hadn't met too many Aboriginals apart from Joe-boy and old Hughie who helped us take the cattle to Charleville. The young men had their packed lunch with them and they all had a chat over lunch and afterwards went on their way. We met these two young men again over the years as we drove in the area. Many years later, in the early 1980s, I met one of these men, Kevin Weatherall from Brewarrina, when I was cooking for Ivan Letchford's shearing team out of Sydney.

We were in the vicinity of Pokataroo when we heard a droving friend of the family, Ray McMillan, had died of a heart attack. Mr McMillan was a lovely man and he still drove with a rubber-wheeled wagon. He refused to become one of the new style drovers who had a truck. Ray had been droving with his son Arthur and another helper. Arthur was riding back to help his father poke the sheep along and noticed him sitting up against a tree, not moving. Sadly he was dead. Six months later, Arthur drowned trying to get his horses off a flooded island when the Mehi River, a subsidiary of the Barwon, overflowed and left the horses stranded there. He was worried the horses would be washed away and drowned if he didn't rescue them. Unfortunately, they survived and he didn't.

Dad had a mare called Elsie who was a bit flighty and as he was riding along one day she shied at a wombat ambling along. This unexpected lunge threw Dad onto a log lying on the ground and he hurt his back. He hobbled around for a long while and eventually improved but the old injury quite often came back to haunt him. After a rough night of sex, he would get around like an old man full of arthritis, calling it "shagger's back". We kids pretended we did not know what he meant, but we knew what caused the noise and the truck shaking if the caravan was still hooked up to it.

One hot mid-summer's day we arrived at our midday stop, and as we were going to be there for several hours, we let the dogs out of their crates and tied them up in the shade. When it came time to put them back into their crates, we found that one of them had been inadvertently left near a bull ant nest and had been bitten so many times he was dead. Ants were crawling all over the poor thing. None of us heard him yelping. Possibly he died from the first bite. It was lucky for Mike, the one who had tied him there, that he was not one of the "good" dogs. Mike had to be punished so he had the option of no tea or a flogging – he chose a flogging. He was a tough kid. The bull ant nests are very small and hard to see if you do not have a good look for them. We had up to twelve dogs at any one time and each stockman had one or two good dogs. The rest were carried for "just in case". Sometimes Dad would take one of the other dogs to keep them up to scratch, but mostly they just got fed and watered along with the others. Drovers had dogs that they called "pan lickers". They were kept for times of extreme heat, when the burrs were plentiful and the good dogs got lame or the stock were needing extra work.

Most of the dogs were great barkers and they used to drive us crazy at times. As soon as the truck started up when they were in their crates underneath, they would start barking and most would keep it up all the way to the next stop. What a sound they made. Some of them fought with each other in their crates so those dogs had to be tied tightly to the crate and we would even muzzle them in extreme cases. It was not unusual to have up to four dogs in each crate tied up. Sadly, once or twice a year a dog was killed in the crate, because we did not get to them in time or we did not hear the fight.

Dad was offered a cattle pup that he called Paddy. Paddy grew into a large cattle dog with the biggest feet I had ever seen. We were getting the dogs out of the crate at the campsite one afternoon and as Paddy got out he spied a kangaroo and took off after it. I had hold of the chain and before I could react, the chain tightened smartly and just as smartly I was on my belly being pulled by the rushing dog. I hung on for dear life and he dragged me along behind him. Beryl was standing between us and the roo holding her doll. I tried to roll over so the chain would miss her but as Paddy flew past, the roo changed direction. The chain caught Beryl on the legs and tripped her. Dolly flew in one direction and Beryl the other. I clung onto the chain but the roo was quickly out of sight and Paddy slowed down to stop. My dress was ripped to pieces and I had cuts and grazes all along my front from the great tussocks of dead grass. Paddy walked back and gave me to lick as if to apologise. Behind me I could hear Beryl screaming. Mum went crook at me, she said I should have let the dog chase the roo. She had forgotten, a week or so earlier Paddy had spied a roo over a fence. He took off and Mike had him on the chain, Mike managed to run, wobbling behind him, trying to keep his balance. The four strand barbed wire fence loomed ahead and Paddy jumped through it and kept running. Mike quickly let the chain go as he did not fancy being sliced with the barb wire. Paddy did not come back until late that night, all cut and bloodied. We did not know if the injuries were from the fences or if he had caught up with the roo and been attacked. Roos will defend themselves if they are cornered and cannot get away. Using their tail as balance, they bring their back feet up and try and rip the opponent's belly open. Because Mike let the dog go, Mum was annoyed with him as she thought Mike should have been able to hold onto Paddy.

We were camped by the side of the road when Dad and the men came in to have their lunch. Dad always opted to have his lunch last so he could have a camp afterwards if the stock were not travelling. We were on the plains with no trees or bushes to use as toilets. Mum moaned to Dad that she could not go for a pee in privacy as the workmen would see her, whether she went behind the truck or caravan. He very smartly told her to piss in his hat, so she grabbed his hat, promptly took it into the caravan and much to our delight and Dad's disgust, she peed in his hat. His sense of humour stepped in and he saw the funny side of it as we were all screaming with laughter. He wore that hat for years afterwards. That was one of the few times that Mum won. At times like this we normally would sit behind one of the truck tyres or use a bucket in the back of the truck but if the stock were spread out all around the camp then yes, someone would see from whichever angle we chose to squat down. The back of the truck with an old tarp across the back door was the only discreet place.

Sometime before this, when I was five and Mike four, Mum said to us, 'I'm going around the other side of the truck and don't you kids look.'

Well hello: what were we not to see? As we heard the sound of her peeing, we looked at each other and knelt down and peered under the truck. Mike and I pulled a face and never spoke of it again. We would not have looked if she had said, 'I am going for a pee,' which was a normal part of our life but with her comment, she made an ordinary event into a mystery.

Les could not resist annoying or teasing his horse, Captain, with flank ropes or sticks. One day he was leaning back and poking a stick into Captain's flank and he pig-rooted. Les did it again and once again the pony pig-rooted and lashed out with his back feet. Les hiked himself over the back of the pony and slid down over his rump and as he hit the ground he again jabbed hard into the pony's flank. Captain kicked out with both feet and connected with Les' face, breaking his nose. Les flatly refused to go to the doctor or hospital and he had a crooked nose for the rest of his life. One of his nostrils never worked properly again and he constantly sniffled, much to our annoyance. His face ended up badly scarred but he never learnt from this experience.

A few weeks later, we were near Trangie and Les put a rope right under Captain's flank. He had the reins in his right hand and with his left hand pulled the rope really hard, Captain bucked and threw him and Les landed on the edge of a culvert, once again smashing his face. He ended up in hospital after this incident as Mum dared not stitch him up as the cut was very jagged and dirty. Dad had to take him to Dubbo, being the nearest large hospital. Dad had family in Dubbo, so after settling Les in hospital, he stayed the night with his brother Fred. The next morning he called in to visit Les and get his prognosis. He then drove back to the camp leaving Les in hospital. As we were so close to Dubbo, every other day or so Dad, Mum and Beryl would drive in to visit him. Les would not have been thrown if he had been sitting properly on his mighty steed but he ruined any horse he rode for any length of time. Les gave the pony a flogging when he returned to the camp. Our sympathy was with the horse.

Les knew no boundary when it came to cruelty and the parents never chastised him over his meanness. One time a drover called into camp and a dog followed that did not belong to him. The dog seemed useless so Les caught it, tied one of our tin rattles to its tail and hit it with a green stick. The dog raced off into the bush with the tin bouncing all around. I was horrified but the men and the rest of the kids thought it a huge joke. For weeks after I worried over this dog was it caught in a fence and starved to death? Did it get stuck on a log with the same result? Or did it find a new caring home? I was a sensitive child and hated meanness and ill treatment to man or beast.

Col on top, Beryl, Patsy, Mary and Mike on the bank of
the Wallam River, Bollon, Queensland

Patsy, Mary & Beryl by the Wallam River, Bollon, Queensland

Emmie & Patsy Kemp in front of Kemp caravan, 1964

Beryl (Mum), Mary & Patsy Kemp

Kemp family at Moree show, 1960

Mary and Patsy Kemp, 1964

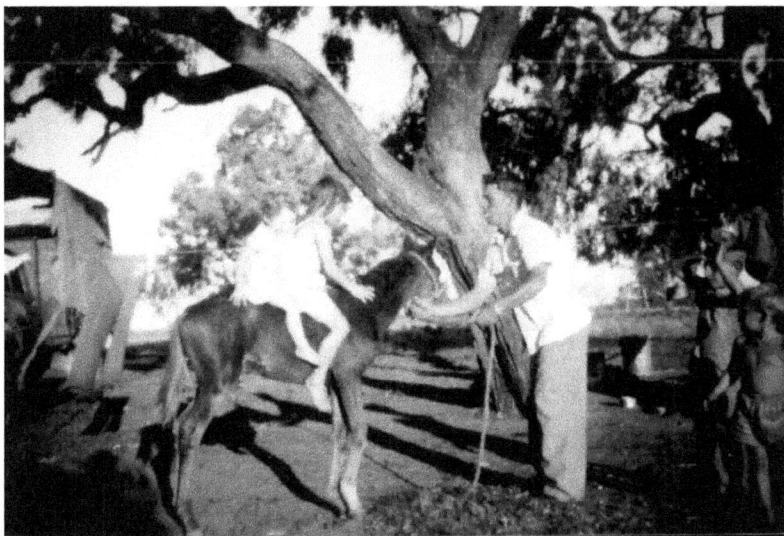

Kids on foal, Mick holding the reins. The truck and clothesline can be seen in the background

From top left: Col, Nanna (Kemp-Thomas), Emmie, Mick & Mary. Front: Mike, Beryl, Patsy, Les kneeling, March 1963

Patsy with Helen on horse, 1964

Mick and Col drenching 7,000 sheep

Beryl (Mum) & Mike taking weight out of caravan,
ready for Mick & Col to pull it back onto its wheels

ST. GEORGE 1958

Traditional land of the Mandandanji people

In 1958 we worked for a stock and station agent in Moree, Ron Hunter, the manager of Winchcombe Carsons. He offered us a job in north Queensland, past Cunnamulla verging south towards Quilpie. We were to pick up a mob of sheep and bring them to the Meandarra sales. As Ron had bought the sheep himself, he wanted the stock to be in good condition upon arrival at the sales. Ron arranged for us to get a stock permit that allowed us to go slowly along the stock route thus helping to keep them in good condition.

When we were on the road for a few days, Dad, Col and Les, and the stockmen left early with the sheep and Mum and us kids had to pack up the camp. This meant putting everything in the truck and caravan, placing the dogs in their crates and pulling down the sheep break.

Mum had been driving for a short while one morning when she saw smoke billowing from under the truck. She quickly stopped the vehicle, Emmie grabbed Beryl and the three of them jumped out of the cabin with Mum yelling for Mary, Mike and I to get out of the back of the truck. We all scrambled to the exit thinking the truck was going to blow up! Mum and Mary quickly unhooked the van while they put the dolly wheel down and pushed the caravan far enough away from the truck for it to not catch fire. Mum warily watched the bonnet, expecting flames to come billowing forth but thankfully it never eventuated and eventually the smoke dispersed.

Mum flagged down a passing car and asked the driver to send Dad back to us as we had an emergency. It did not take long to see Dad cantering towards us, a worried frown creasing his face. After a look under the truck and in the cabin he discovered that she had left the hand brake on which had caused the smoke. No damage had been done, so he took over driving to meet up with the stock. As they drove along, he gave Mum a lecture about the wisdom of taking the hand brake off before moving the truck too far.

Mary had the enviable job of cantering the mare to the mob a couple of miles away by now. Honey-bee was probably one of Dad's best deals as she was a tall, beautiful chestnut filly. Just the slightest touch on the rein would get her moving and turning, a slight kick and she would gently canter and apart from the prancing all the time, she was a pleasure to ride – but she never actually stopped! I was once cantering on her and did not see the bore drain ahead. Honey-bee jumped over, her head swung back, hitting my forehead with a mighty bang. My head and neck hurt for ages afterwards. Whenever you removed Honey-bee's saddle, she would shake herself, lie down and wriggle her back in the dirt or grass and if any water was nearby she would roll in that. She did this every time, the only horse I have seen do this on a regular basis. She would try and roll in a river and it did not matter if you were still on her. Once she started to paw at water you knew it was time for a swift kick in the ribs to get her away from temptation. She certainly kept me on my toes. I would much rather give a few extra gee ups and a kick in the horse's ribs to a quieter horse than have a horse in flight mode all the time.

We caught up with the sheep and Dad continued past the herd for about half a mile. The fire was lit, the billy put on the three pronged-tripod over the fire to boil, butter, Sao biscuits, tomatoes, cheese, Vegemite were all plonked on the table and *voila!* morning tea was ready. Dad was peckish, he sent Mike to retrieve a shovel from the back of the truck, plonked it on the fire and used it to grill some chops. The fat was cut off and grilled to a crisp giving off a mouth-watering smoky smell. These lamb chops were delicious on top of a slice of bread and our meat was always medium to rare.

When we were killing sheep in summer they had to be killed late in the evening as the flies and blow flies were unbearable. After a sheep was killed, it was fully wrapped in a sheet to keep the pesky flies off the carcass. On the side of the truck there was a permanent pulley that pulled the sheep up so it was easier to skin and gut. After it was "dressed", the sheep was pulled up even higher out of reach of the dogs. If a dog got off its chain it would jump up and try to take bites out of the neck hanging down. When the dogs did this, it made the truck rock and it would possibly wake one of us up. If we could be bothered, whoever woke up would get up and tie the offending dog back up but this rarely happened, we would be either too tired or lazy to bother.

In winter, there were not as many flies around so it was okay to kill the sheep late in the afternoon. Dad would cut the necks up and we had them in stews and curries. Before it got to that stage though, the skin was kept on the carcass for as long as possible so the sheep was kept clean.

Killing and skinning was done on the bare ground. A slit was placed above the back leg hock to put the gamble through each one and that was used to pull the sheep up to the desired level, then the rope was tied off at the side of the truck. On a kill night the brains, kidneys, liver and heart were cooked, smothered in onion gravy and we normally just had bread with it. The family hierarchy ate their favourites, Dad always had the brains, Emmie the heart, Beryl the kidneys and the rest of us ate what was left. Most nights if the main meal was a bit lean, we had a pudding, be it custard and tinned fruit, a jelly, junket, White Wings puddings or sago. We were not too fond of sago pudding, saying it was frogs' eyes. The sheep's head was cut in two and the dogs had half each and the guts. Jelly and junket were not made on really hot evenings as it was too hot for them to set.

In summer when it was really hot, part of each sheep killed was salted with coarse salt and put in a sugar bag and hung from the side of the truck. To get the salt to do a good job, the meat had to have large slits in it and the salt was packed into those slits to get the meat nice and dry. As it dried, the meat dripped wet and salt, and you had to be careful if you brushed against the bag as it did not feel so good. Salt did not necessarily mean the meat did not get fly blown. If it happened mostly it was discovered just before cooking, so along with the excess salt, the maggots were washed off and the meat was boiled as normal. If it was fly blown, we just did not discuss it: life went on and only those who needed to know, were told. Usually after telling Mum she would say, 'A few bloody maggots won't kill ya.'

I drew a line if the meat was actually rotten as you could always cut the rotten hunk off and throw it to the dogs – drovers' dogs are always hungry! This salted meat tasted much like corned meat today and it could only get back to an edible stage by being boiled. If the meat had really dried out it had to have a couple of changes of boiling water to keep the salt down to an edible level.

Dad used to get out of bed when Mum got up, just before daylight, as the sheep had to be cut up into cooking portions. We had a tomahawk that was kept specifically for cutting up the killer and a chopping block was a piece of cut off log about twelve inches high and eight inches across. It was only used for this purpose – never as a seat.

We had roast legs of lamb quite often for our meals at night. We were connoisseurs of a roast of any kind and I have never understood someone not being able to cook a roast. The roasts were always cooked in the camp oven over the open fire and the older the fat the tastier the meat. Once it was nearly cooked we would place the potatoes and pumpkin in the same pot and the vegetables would come out brown and tasty. Mum always made a delicious gravy and we had carrots and a tin of peas or beans. We used the shoulders of the lamb to feed us on alternate nights. They were mostly boiled in salted water and the vegetables peeled, cut up and plonked in the pot to cook and that was our "tea" at night. The chops were used for the midday meal if we had no cold meat or, on occasion, morning smoko as it was a rarity to have an afternoon smoko break.

If there was not enough meat for a night meal, a stew made with bully beef was on the menu. The stew could be turned into a curry with the addition of several spoons of Keens Curry. Sometimes Mum would smarten up the curry with a handful of raisins but most of us kids did not like the "dead flies" as we called them, ruining our good tucker. We much preferred to eat the raisins as a sweet and that was a rare treat!

If there was no cold meat for lunch, a couple of cans of bully beef were opened. We thought this was horse or kangaroo meat as it was so stringy, but I liked it because it was different and quite salty and greasy. Dad enjoyed a tin of braised heated up and served on toast with lashings of PMU sauce.

As most old drovers only ate corned meat and damper, they could suffer from Barcoo rot (an ulcerous skin infection). Once they got to the nearest town they would buy pineapples and eat as many as they could. If not pineapple season they ate oranges, limes and lemons. Some would boil pig weed and they would chew on it as it has a slight salty taste and is quite palatable. We kids would nibble on it just for the saltiness, not that salt was held back from us at the table.

Our first job on arriving at a new campsite was to clean the area before unpacking the truck for the night. We would look for any "good" stuff that other drovers had left behind. Cleaning up someone else's campsite was a treat not a punishment as we found our treasures there: discarded books, broken tools, misplaced items and coins and even the odd pound note! We rarely found toys as very few drovers had their families with them. Books and magazines were quite often left deliberately where they would not get too damaged so whoever followed could have a read too. A lot of the books had the last few pages missing – the drover would have taken these pages out for toilet paper! I read them anyway and then would think about possible endings, which was sometimes very frustrating. Then the truck could be unpacked, the sheep break put up, dogs tied, watered and fed and we could have some time to run around. School work was never our priority. If Mum did not mention it, neither did we.

Late one afternoon, Dad spied some nice fat lambs in the paddock next to us and told Col to catch one of the sheep for a killer. Something went wrong with his trusty dog and stalking, so Col ended up doing a lot of running and in doing that he lost his balaclava. For ages afterwards, there was always the worry about that clothing item being found and traced back to us. But with the size of those paddocks, who would chance upon a balaclava with no one knowing who wore it? Col was banned from wearing a balaclava when on these shifty manoeuvrers after this incident.

Col and Les were great in the saddle. Some of the camp horses were so toey that the rider could barely touch the stirrup before the horse would either shy or buck away. Col and Les would just hang on like monkeys until they managed to get their legs over the saddle and into the stirrups. Once they got their bums firmly on the saddle they would spur the horses in the ribs and give them a hearty gallop to steady them down.

We girls in the camp wore shorts or a dress and if a bit of horse work came along, we rode in what we had on. I tucked my skirt under my thighs to stop the flapping skirt from frightening the horse. Experience taught me a flapping skirt on a fresh horse was a recipe for an interesting and uncalled for ride! There were no long pants for the camp kids though, except for Beryl. Interestingly, we rarely had our thighs or calves red raw from riding bare legged so I guess our legs had toughened up over time. After a couple of spills from a flighty horse I refused to ride them again so I only rode Dad's horse or a quiet pony or horse that was guaranteed not to buck. I did not mind the falls but I dreaded the dead stop as I hit the ground.

Mary rode in her dress with the skirt flapping in a very cavalier manner with her red curly hair flying in the wind. My legs were never long enough to fit my feet into the stirrups, so I either rode with my legs just dangling down or I put my feet in the join of the leathers where they buckled up. With my feet dangling it got a bit painful on my back after a while, so I would take it in turns putting a leg over the pommel to ease the pressure. Whenever the others saw me do this they made me put my feet down as sitting like that was a recipe for disaster.

I loved horses and loved to ride as often as I could but I didn't get to ride much. It was only in scrubby country or if the men were in a hurry and a couple of the stockmen had to come in at once that I would get to relieve them of their duty over the dinner break. Usually they ate their meal, had a cup of tea and were back out again in maybe thirty minutes maximum. Other than that, as the mob poked along it was quite achievable for the men to come in one at a time. Even then, it was only four men at the most with the mob. If there was no fence line, one stockman was on each side of the mob and one in the front and Dad in the dust behind them as they grazed along.

Sometimes when I was out with the stock, I would deliberately not look back at the camp to see if I was being called in. I would hear Dad whistle and cooee to me but I would ignore him. Speck, his dog, would look at me as if to say, 'Your father wants his horse back!' But as Speck ignored all my requests to work for me, I didn't see why I should have to acknowledge any of his requests. One or the other of the men would shout out to me and I would shout back, 'I can't hear you.' If I heard Dad was getting annoyed I would reluctantly go back to camp but if he sounded ok I would keep riding along until he came up to the mob driving the truck with Mum in the passenger seat. He would pull me off the horse and give my bottom a light smack. He could not begrudge me tailing along behind the mob enjoying the ride.

One particular afternoon we drove into our campsite early enough for us to set up and still have time to do some schooling before tea. Mum called for Les to come and join us. Nearby were road works and being the weekend, the men had gone home. The workers had left a pile of 44 gallon drums that had been emptied of tar on the side of the road. Drums were great for kids to have fun in. Most of them were on their sides and some were standing up with the tops cut off.

Les ignored Mum's request to help us kids get the dogs out of the crates and he started to run over the drums. He soon lost his balance, slipped and cut his shin quite badly. Upon landing on a drum his bottom lip was badly cut much like mine some years before but his gash was much larger. Mum placed a clean tea towel over the bottom part of his face to try and stop the bleeding and she drove Les, Beryl and Mary back to the stock a couple of miles away. Mary swapped places on Dad's horse as the men needed the extra worker and Dad dropped Mum off at the campsite. He drove Les into the St. George Hospital where he stayed for a week until he was capable of walking, so another stint in hospital for Les. His face scarred much worse than mine as the tops of the drums were much larger than the kero tin I had fallen on. He didn't lose his teeth like I did, though he always had crooked teeth.

While on this trip we met up again with Roy and Marion Wilson and the kids, Barbara, Judy, Barry and Peter. The adults were having a beer one night and Mum had a drink with them. Mum was referred to as a "one pot screamer" as after just one drink she would end up jumping on the table and doing the "highland fling", skirt going every which way. Dad laughed and clapped along with the rest of them but not long after this, she stopped having a drink and begrudged any other woman from having one also. The Wilsons were good fun and we all got along fine. On occasion we would camp together so everyone could have a catch up. In the light of the fire the adults would sit around telling dirty jokes and we kids would be sent to bed but sometimes we would sneak back out again to listen to the jokes and chatter. If the grass was tall we would lie in it so we could not be seen and there was always camp gear lying around to hide behind. As the truck and caravan were not far away we could quite clearly hear the jokes. Some we did not understand but some we did and we managed to re-tell them to the others who had not heard. It got embarrassing if one or the other of us fell asleep while hiding and could not be woken. The unlucky one would get tripped over in the dark or would wake up at daylight wondering what had happened.

Sometimes we were allowed to stay up but if we misbehaved we were sent to bed as punishment. If this happened, quite often we would sneak out again and slink at the edge of the campfire in the darkness, listening to the goings on of the family. One night Col was a bit mouthy so he was banned from the fire. We needed more wood and I was told to go and get some. I was scared of the dark and who or what was lurking out there so I reluctantly walked to a large gum tree nearby that had plenty of dead branches lying under it. As I walked in the dark this big ungainly creature approached and I screamed, frozen on the spot. It was Col with a large load of wood in his arms. I cried and laughed at the same time, crying because my worst nightmare had just been realised and laughing because it was a friendly one.

Another night I was gathering wood in the moonlight under the gum trees and as I leant down to pick up a piece of nice looking wood, it slithered away from me. That gave me the creeps and I have had a fear of snakes ever since. Dad killed a snake in the camp one afternoon and made me pick it up by the tail and carry it away from the campsite. Even now, fifty years later, I can still feel that squirming snake!

The Wilson boys showed Col and Les how to make shanghais, Les soon found another way to torture animals. Mary and I loved to have a go with the shanghai too but we only aimed at the base of trees or other targets. None of us girls got any satisfaction out of animal cruelty. The shanghais were made with a medium size forked stick, cut to suit hand sizes. Tube from a discarded tyre was cut to size and tied to each side of the forked stick. You then put a small stone in the centre, pulled it back to your chin, aimed and fired. Mary and I were never good at it but we persevered, trying to beat the boys.

Roy Wilson had a .22 rifle for sale and Col bought it. We all learnt how to shoot and we found it an enjoyable pastime, shooting at tins or bottles. It was kept under the front seat of the truck and we were all free to grab it and have a bit of practice, though Mum was never pleased, thinking we would shoot each other.

The Wilson's had a rack under their truck and they had six chooks that laid eggs on a regular basis. We were fascinated to see these chooks jump up under the truck and lay an egg then come out cackling, boasting about their great deed. It was not a regular sight with droving teams. The senior Wilson was a large family and they were Crow (Roy), George, Gunna (Fred), Spot (Charles), Sister (Margaret), Tiger (Lindsay), Whitey (Leslie), Darkie (Edgar) plus others, 17 siblings altogether. Darkie was travelling to work one day and travelling rather fast on a gravel road, the gravel gripped one of the tyres and the car flew into a tree. The car's engine ended up several feet up the tree, Darkie was not badly hurt but his pride was. From that day onwards the story was he had accidently put goanna oil in the engine and that was why the car went partway up the tree.

Mum had quite a few funny sayings. If she was resting on the bed in the caravan and we annoyed her by asking questions she would say, 'Can't you kids think of another word except "Mum"?' She hated to be interrupted while she read and smoked, which she did a lot in the early years. If we were restless and fidgeted, it made the caravan rock and she would snarl, 'Stop that fiddle farting around will ya. You kids are fidgety, like a fart in a pickle bottle.'

I think of that comment every time I see a bottle of yellow pickles and it still brings a smile to my face.

MEANDARRA 1963

Traditional land of the Burrungum or Jarrawah people

Col was now seventeen, Emmie sixteen, Mary fifteen, Les fourteen, I was thirteen, Mike eleven and Beryl eight. Emmie was a lovely looking young girl with her white skin and long flowing auburn hair and Dad decided as the hired workmen were starting to ogle her, they had to go and she and Les would now take their place. Before this stage, Les had been with the stock but not considered a stockman as he would canter off at will and do his own thing, no doubt causing grief to any local wildlife. Marsupials were a target for him and his dog Snip.

Now that Emmie became a stockman, Mary had the job of breaking the camp up and packing everything away while Mike and I gathered up the dogs and placed them in the crates straight after breakfast before helping Mum pull down the sheep break. This entailed first pulling the iron pegs out of the ground and placing them on the ground beside the fence as it sagged over. Sometimes a peg would be hard to pull out so the hammer came into play and we had to hit the bottom of the peg a few times with the hammer to loosen the soil around it. After all the pegs were pulled out, either Mum or Mary rolled the wire up into three rolls. We had to lift the wire and tighten it because the rolls had to be a certain size to handle and put in the back of the truck. We carried the pegs back to the truck and as there were quite a few, there were lots of trips. On hot days the pegs were hot to carry and in the winter they were freezing, so we kept the fire going to quickly warm our hands ready for the next load. The pegs were also rusted and dirty. Weaklings like Mike and I could only carry three at a time as they quickly became quite awkward and heavy.

Beryl never got out of bed until everything was packed away and we were ready to drive off. She would have a bowl of Weeties standing near the dying fire as we did the final bit of tidying up around the camp. No matter how bare it was around the fire, we always poured the final bit of billy tea on the fire and then the wash water. If the fire was still smouldering, a shovel or three of dirt was thrown on to smother it completely. We didn't ever know when a wind would gust up and spread the fire coals or a willy willy would merrily scatter them into any dead leaves or grass nearby.

After joining forces with Winchcombe Carsons in Moree and Bruce Knobbs in Meandarra, Dad knew his worth and started to charge accordingly for his services. With these agents, he began to make more money, not that we kids saw any of it. Sometimes Dad would talk about what HE would do when HE became a millionaire. It was always HE who would become the millionaire not US. He had given up heavy drinking and started to buy lottery tickets. He would buy £10 worth of tickets while we ran around with the arse out of our pants and no shoes, hats or treats. When he was checking the results and could see he hadn't won he would say in disgust, 'If it was raining quids I'd pick up a bloody summons.'

Our big break came when we had to travel to Bourke, western

New South Wales to pick up over 7,000 head of sheep and take them to Meandarra Queensland.

Sid Derrick, who owned a sheep station called *41*, with Winchcombe Carsons' backing, went to the Bourke sales with Bruce Knobbs. The auctioneer stood in a pen and auctioned off 1800 head. He said to Sid, 'You can buy by the pen or the whole lot if you want!'

Sid retorted, 'I will have them all!' So he bought out the whole 7,000 in the sale yard in one bid. A few other bidders went home disappointed that day but Sid was elated with his deal.

It was mid-summer and as hot as Hades, well over 100°F (38°C) in the shade. The sheep were counted out of the yards, Dad signed the paperwork and we started off on the road heading towards Moree. His first job was to drive into the Bourke Township and get a travel permit from the police station, which gave him the right to travel on the stock routes. These permits carried a copy of the ear mark and brand of the sheep that he was in charge of. The whole mob was from one station further out west and they all had the same brand and ear mark. This gives you an idea of how big the stations out west were.

Mum started the truck and drove three miles ahead of the sheep for the dinner break and the heat was intense and relentless. Inside the truck and caravan it was like an oven. There was no breeze or shady trees in sight, so we took the dogs out of their crates as they would quickly become overheated all being locked up together in such a confined space. They were tied up around the truck and given a drink. I placed a potato bag under the truck because it was too uncomfortable in the caravan, and lay down with the dogs. It was probably a degree or so cooler there but the heat emanating from the dogs did not help. One crawled over to get on as much of the bag as it could. Mike decided I looked comfortable so he soon joined me with his own bag and eventually we fell into a heat-induced doze.

As the stock got closer, Mum called for me to get a loaf of bread out of the bread tin in the back of the truck. This tin was a large galvanized rubbish bin used to keep the bread fresh and the mice away. On this occasion, I opened the lid and saw the bread was green with mould. This did not deter Mum as she cut the mildew off, sliced it up and plonked it on the table. I complained to her the bread was all wet and had hairs on it and she replied tersely, 'Eat the bloody stuff or go without, you're lucky to get that.'

As I bit into the bread and pulled it away from my mouth, the string from the mould stretched out and it then became a game to see who could make the longest string without breaking it.

That night upon getting into camp, Mum made two dampers and two brownies. It was a big ask in the extreme heat as they had to be cooked over an open fire. The following day at midday she made a batch of scones cooked in fat and we found these delicious. Sometimes for a real treat Mum would add dates or mixed fruit. She only cooked a damper in an extreme emergency so we knew we were getting low on bread if a damper and or a brownie appeared. If we camped for a period of time she would whip up lamingtons, which were delicious. The icing was just right and she did not stint on the coconut. There were always plenty of volunteers to lick the bowls out, little fingers and tongues lapping up the treat. On occasion, if Mum was in a really good mood, she would deliberately make extra icing so there was more in the bowl for us to share.

Ants and flies were our biggest problem at meal times. It was quite common for them to get into the tins of jam, peanut butter, syrup or treacle if the lids were left open. It was common to hear the cry, 'Who left the lid off the syrup?'

'Not me,' was the answer. No one ever owned up to being the last at the table. Mum often made the comment that, 'Mr Nobody had better start to pull his weight around here.'

Mr Nobody was the tenth family member and he never left home! No one ever quibbled over eating this spoilt food. If there was any sort of complaint, Mum would snap, 'Eat the bloody stuff, a bit of fly shit won't hurt you.'

There were not many jars around in those days so the jam's tin lid was pushed back over and an old rag or a tea towel would cover it to keep the dirt out. If we were having a meal in the caravan and flies got in, Mum would grab up the old tin fly sprayer and start pumping at the flies, the mist going over us and the food. We chewed on regardless; we all had stomachs as tough as old boots!

Eventually the sheep caught up with us and the men took turns to come in for lunch. Dad came in first and Mary rode his horse out to the stock to let Les come in. As each one had their lunch they went back out to the stock to relieve the others. Dad always stayed in the caravan for a rest while Mary and I relieved the workers.

If we were in a hurry, the dirty dishes were not washed. They were put in a dish on the floor of the caravan and carried to the night break, but as we did not have enough dishes to do two meals, they had to be washed before having the next.

We learnt to rub the bottom of the billy on the ground before putting it in the caravan. One time we put the billy into the van on top of a potato bag to stop the black from getting onto the floor and when we pulled up at the next stop Mike smelt burning. Upon opening the caravan door the bag was smouldering getting to the point of catching fire. Fortunately for Mike's sense of smell there wasn't too much damage to the floor. The van could have caught fire if we had travelled much further.

Normally when we were down to our last couple of loaves of bread, Dad would flag down a passing car and ask them to get us a dozen loaves in town and drop them off on their way back. This worked fairly well and only occasionally did the drivers refuse, usually because they were not returning that way. Dad would give them a quid to pay for the bread and rarely did he lose his money. Quite often we did not hear the car pull up and drop the bread into the cabin of the truck. We were all good sleepers. The dogs would go ballistic but because we were so used to hearing them bark for no reason in the middle of the night, we weren't disturbed by them. If we stayed on the road long enough, the same people would pull over and offer to get the bread for us again and sometimes they would drop off a bag of lollies for us. On occasions, the wife would bake a cake for us. Most people were very kind, considering we had taken stock through their property and we probably had two or three of their fat lambs in the mob for killers.

There was a hierarchy of who sat in the front of the truck with

Mum and Beryl. Before Emmie went out with the stock, she was always in the front but now there was an extra seat that was graciously handed out to whoever was in the good books at the time with Mum or Beryl. Beryl would queenly hand out her favours to the favourite of the day. Mary and Beryl got along famously but Mary and Mum did not so that created a bit of a problem. I was not fussed about sitting in the front of the truck, much preferring to have a good book in the back of the truck and if I had an audience, which I often did, I would read out any interesting bits. We often had cowboy books because we swapped with the stockmen. *Larry and Stretch* was a favourite as they were easy to read – no big words. Mind you, I was not a great reader then and I had to put bits in where I thought they fitted, as I did not know how to read certain words. We all had a go in the front of the truck, but possibly not every week as I liked my reading time and read the same books over and over. I was a helpful girl and enjoyed pleasing the others but I could also be stubborn – I took after my Mum! It never seemed to worry Mike where he sat as he was the most obliging of us all. If you wanted to share a job, he would help; if you asked him to do it, he would do it for you. He was certainly my favourite brother.

When travelling on private property, the stock was kept in paddocks and on the road there were gates to open and close. The men had to do this to stop the sheep or cattle in the adjoining paddocks from getting mixed up with our stock. Sometimes Dad had to drive ahead of the stock checking out the route for the next week or to see how the feed and water situation was. On these trips he would choose the next few nights' camping spots. He had an excellent memory and could remember where the water was and where the best stock feed and camping spots were.

After one of these trips he came home laughing and told this story. As he was driving along the road there was a chook standing on the side of it. As the truck got near the chook started to run very fast, keeping up with the truck's speed. Dad noticed it had three legs and he then speeded up and the chook still kept pace with the truck. Dad once again topped up the speed and the chook kept pace.

Suddenly the chook turned left into a station's road and kept running. Dad was intrigued and stopped the truck, reversed back to the side road and drove down to the house. A man walked out to greet him and after a short chat Dad asked about the three legged chook. The farmer scratched his head and said, 'Well me, the wife and son all like drumsticks so I decided to breed three legged chooks.'

Dad was impressed. 'How do they taste?' he asked.

The farmer replied, 'Dunno, haven't been able to catch one yet.'

We enjoyed this story and for years we kept our eyes open for three legged chooks. We were well into our teens before we realised it was a joke.

Dad called into a station to get the three 44-gallon water drums topped up with water and the manager he spoke to refused, saying he did not pander to the drovers. Dad then called into the neighbouring station and the owner there agreed to give him water. Dad told him that his neighbour had refused and the owner said to Dad, 'If a shark bit that bastard he would be too miserly to give a shout.'

Most station owners and managers were only too happy to help out the drover, knowing they did a good job when moving stock.

Quite often just by chatting to station owners on the way, Dad found out how the land was ahead for the mob. If the water in the tanks on the stock route was not suitable and we got low in water, he would have to take the truck into a property to ask if he could fill up the tanks with clean water. It was unheard of to be refused drinking water and Dad would disappear into the house of the friendly owner or manager. He always had one of us kids with him to fill the drums with fresh water and quite often the lady of the house would come out for a chat.

The girls' bed (the one I shared with Emmie and Mary) was only about a foot above the drums and the two single beds for Les and Mike ran parallel to the truck. At night we could hear mice running around the rims of the water and fuel drums, squeaking and playing chasey. They also ran around the rest of the truck and in the crate over the cabin. We would get the giggles listening to them when we were trying to fall asleep. On occasion, if they bred too much, we had to do a mouse cull. We all had fun as we scrambled around looking in the hidey holes and finding nests full of baby mice. We then fed these destructive little pests to the dogs. This job was never Emmie's, as it was beyond her sensibilities, but Mary, Mike and I never quibbled over the deed. The mice were very destructive to all the gear. They would chew the saddles, bedding, food and anything else they got to, so they had to be kept in control. I was not squeamish about doing away with these cute little beggars that could make our life miserable with their overbreeding and the yucky, musky smell of their pee.

When we took charge of the sheep, they had about six month's wool on them. As the wool got longer, blowflies would strike any wet, damp area on the sheep, mainly at their rear ends or the wethers' pissles. When a fly-stricken sheep was spied, this would entail jumping off the horse, rushing into the mob to catch the sheep, rolling it over and placing it into shearing mode and cutting away all the affected areas. This included the wet area and a certain amount of dry area and any dags that hung down. All the workers had a pouch hanging off their saddle that carried a pre-mixed bottle of KFM (Kills Flies and Maggots) and a pair of sharp hand shears. After the sheep was dealt with, the pre-mixed KFM was then liberally washed over the wound area to kill both maggots and keep away any flies for the next few days until the wound dried and healed. In the summer it was an ongoing job to catch and fix any affected sheep. A major sign a sheep was fly-blown was that it would keep to the back of the mob and quite often lay down. This was a sure sign of the mob needing to be crutched in the near future or if full woolled, needing to be sheared.

Sheep dip is poisonous and when mixed it goes a white milky colour. One day Dad refused to let Mike go with him on a trip into town. Mike got really angry with Dad and threatened to poison us all.

'Well, how are you going to do it?' we asked.

He replied, 'With the KFM.'

We all thought this was a real joke and for years afterwards he would threaten us with, 'I am Mr KFM man you know.' He was never taken seriously and he always got a good laugh. Sometimes when we were having our Weetbix for breakfast, if the milk was already made in the plastic jug one of us would give it an exaggerated sniff and ask, 'Did YOU make the milk, Mike?'

Les had been complaining for some time that Captain was now too small for him, so Dad bought him a quarter horse that Les called Gaywallah. After getting this horse he refused to do anything else but stock work. Les still had Snip and he spoilt the dog by teaching him to look for and kill lizards and small animals. When needed, Snip was quite likely off looking for his own personal sort of "stock". That dog would not work for anyone but Les. Gaywallah was Les' horse and he bought a hackamore bridle advertised in the *Hoofs and Horns* magazine. It was sad to see how Les treated his horse with that bridle as he would gallop along and just pull back on the bridle so Gaywallah had to practically stop on his haunches. Gaywallah got so skittish that as soon as Les' foot was in the stirrup and his right foot left the ground, the horse galloped off with fear in his eyes and his head up in the air.

One particular time, Les was riding away from the camp after a dinner break and a magpie attacked him. Both he and the horse disappeared into the distance, galloping uncontrollably. These antics worried and stressed Mum but Les didn't worry about anyone's feelings. This type of incident gave him more fuel to do dangerous antics. One of his party tricks was to get his horse into a gallop, reach up and grab a branch and let the horse gallop along riderless. We would have to catch the horse, which would more than likely run into the mob of sheep or go to the mob of horses wherever they were. After a lot of running around we would then take the horse back to Les, if he could be found. He would think it great fun to sit on the branch waiting for you to find him. Dad was very tolerant of Les' antics and the rest of the siblings knew it was a way for him to get attention.

The yearly local show was on in St. George and Dad and Mum decided we would all go. It would be our first agricultural show. Because the camp was well off the road, it was decided that Col did not have to stay behind to "mind" it. Mum had the money so while Dad went to the bar, the rest of us enjoyed the show. We enjoyed the "nock em dollies", firing at the ducks in a line, going on rides and walking into the side shows. It was a long and tiring day but enjoyed by all and we did not know half the things in the show existed. I think we were all sick from the rich sweets we ate but my favourite was the fairy floss. Yum! Les won a china horse and we all coveted it. While we were walking around the show we came across Dad standing on the box to have a go at the show's "Top boxer in Queensland". Mum was horrified to see him there and she tried to talk him out of it but he was determined to earn the £10 offered to knock the boxer down. Eventually Mum decided to go into the tent to watch, leaving us kids with Emmie. We had to stand near a corner until Mum and Dad came out of the tent, Dad covered in blood. No winner was announced to us and Dad staggered back to the bar.

After watching the ring parade of the stock, Mum decided it was time we all went back to the camp. She found Dad drunk at the bar and managed to get him to take us home. The back lights on the truck were not working and we kids in the back had torches with red rags over them with one kid at each side of the truck. When Dad went to turn, we had to flash the torch on and off. Thank goodness we were doing this as we went past a cop directing traffic! Dad was so drunk he swerved from one side of the road to the other but we made the twenty miles back to camp all in one piece. In the early days of droving right up till the early 60s, Dad rarely came back to the camp sober. In between jobs he spent his time in the pubs, as that was where the jobs were picked up until you mated up with a stock and station agent. If you were good at your job, they kept you in work when they could.

Mail day was a great day of expectation and a happy time for us all as it meant a letter or three from various parts of the world. Col had a girl penfriend in the Philippines and we all took turns to write a letter for him. Emmie had her friend Shannon from Dirranbandi, Mary had a boy in Papua New Guinea and I had four penfriends: Kerry Thompson in Sydney, Linda in Swindon England, Kay in Yarraman Queensland and a boy in Ceylon. I was an avid stamp collector and purloined the stamps from the others.

If one stopped writing to us, we would manage to get another one. The Sunday paper, which we rarely got, had a kid's page called *Chuckles* and penpals were listed in it. They were a great part of our lives and we enjoyed getting the mail but not the school correspondence that arrived at the same time. I had my penpals for many years. I am still an avid letter writer and even now my emails are long. My communication skills are still far better by pen than verbally.

Another thing we liked to do was to write into 2VM, Moree and Bourke's radio station, and request a special song to be played for someone we knew. They would even pass on a brief message, and the person you played the song for would do the same for you. After a while the announcer would get used to who was calling who and make a comment or a smart remark.

When we could, we would stop at 1 pm on a Sunday afternoon to listen to the radio and hear our requests played and have a laugh over the smart remark the announcer had made of our message. One of the popular songs we requested was *Judy, Judy, Judy* by John Tillitson and as Emmie had a dog called Judy, she would sit and wag her tail and have a silly grin on her face as she listened to the song. The radio was not used too often as the battery was not cheap and Dad objected to it. We were country and western tragics so Johnny Horton, Hank Williams and Buddy Holly were our favourites along with others. We girls were very fond of Elvis Presley. Mum loved Buddy Holly and would sing along to his songs. Dad did not like music much so when he was around the radio was not turned on unless there was something he wanted to listen to such as the Melbourne Cup, boxing matches and every four years, the Olympics. Another of Dad's "mean streaks" was him slinging off at us because we were not clever like these people who were winning the medals. If only we had half a chance to try! If we did something stupid or the wrong way we would get, 'If yer brains were bloody dynamite it wouldn't blow the hat off yer head.'

Bull ants were our enemy when playing and we were always aware of their nests. They built their nests up with small stones and sticks, and the jumping ant would pop up unexpectedly. The bite from them is very painful. The cure for this was getting a small pinch of tobacco, spitting in it, rolling it into a ball and applying it to the bite. For some reason this did take away a lot of any sting.

One day Mike encouraged a bull ant to get onto the tip of a stick. The ant was balanced and Mike said, 'Here Beryl, grab this.'

She grabbed the end of the stick and also the ant, which proceeded to sting her on the finger. Mike denied knowing the ant was on the stick but he later confessed to me his intention had not been to hurt Beryl who was aged about five at this time. We all had a giggle about it later as we could never get our own back on Beryl, she was always protected by Mum. Beryl was "the baby" from the time she was born until she left home.

As we were now "in the money" and had long months of constant work ahead of us, the parents decided it was cheaper and more convenient to get our groceries in bulk from QPS (Queensland Pastoral Supplies). Groceries were ordered to last several months. Cartons of cheddar cheese, tinned peas, beans, meat, fruit, powdered milk, porridge, Weetbix, Cornflakes and for Beryl, Weeties, tins of syrup, treacle, honey, billy tea and jam, Vegemite, cartons of Sao biscuits, tin boxes of Arnott's sweet biscuits, bags of flour and sugar. Because the order was quite large, we also received a tin of boiled lollies for free. What a treat for sweet starved kids. These lollies were doled out one by one until they were all gone. Mum and Beryl got the bulk of them. If we complained and made an issue of it, we received even less. The order arrived by public transport, mostly trains and I cannot remember ever getting any broken cartons. Sometimes the groceries eventually had weevils, but one had to eat! It was around this time that packaged cereals began having a toy placed in them. In Weeties it was a plastic horse. Amazingly, when placed on a firm item on a slight tilt, the horse would walk down. Later there were various other animals that did the same. It was always a fight as to who was going to open the pack and rummage through for the toy.

Bull head burrs were deadly on both dogs and humans. They have three hard prongs and stick into the flesh and hurt like crazy and when you pulled them out they stung. It was not uncommon to be chiacking around, running from your opponent and end up in a big patch of burrs before you realised it. Then you either had to get someone with shoes on to come and piggyback you out or pull the burrs out and gingerly walk back. This could involve having to rub the top soil clear of any burrs with your bare feet before taking the next step. We thought it was fun to go to offer to help the one in the burr patch and then push them over. Many a time when trying to help, you would end up lying or sitting in the burrs as you lost balance. This would either create gales of laughter or tears of pain, depending on who and how it was done. We would sometimes throw our shoes to the person affected so they could get themselves out of their predicament.

Mum willingly got the burrs out for us. She would sit on a stool and we would lay on our tummies and put our feet on her lap and we would "ooh and ouch" as she did a bit of digging, never too gently. We must have had tough feet as we never seemed to get that many in our feet, but maybe it was so often we did not think anything of it. In return we had to pluck Mum's grey hairs out – we all found it a tedious job except for Beryl who loved doing it and she would lovingly comb Mum's hair and pull it back behind her ears and keep it in place with bobby pins.

When Dad had to go to town, Mum ironed his shirt with a methylated spirit iron. Laying a couple of towels on the table in the caravan, she would light up the iron, which she viewed with great hatred and fear, and wait for it to get hot enough to use. She would first test it with a wet finger, if it sizzled satisfactorily she began doing the hated job. Dad always went to town, sale yards or to the stock and station agents well dressed. The white shirt (never pink now) was always teamed up with a good pair of sports trousers. RM William riding boots, shined with a final spit and polish by Dad himself. His hair was neatly combed with Brylcreem, his teeth were brushed and the final touch was a dash of *Old Spice* aftershave. If I stood close enough to him while he was splashing it on, he would put a dash on my nose, I loved the smell of it. It was the only perfume in the camp.

Being in the country, everyone waved to each other. If we were camped right off the road, friends would blow their horns when passing to make sure we saw them and we would all wave and cooee back and then wonder where they were off to. If it was a vehicle we did not know, then we would chat over who had bought the new vehicle, never thinking it was a stranger just waving to us.

Dad went into Walgett to get some groceries and fill up with fuel and he met up with his sister Anne and her husband Wally Kinsella whom he had not seen for years. They were working in Walgett, fixing and selling sewing machines from a shop. They travelled a good part of New South Wales doing this work. Their base was in Dubbo where Dad and Aunt Anne's mother lived in their house and looked after their teenage daughter. If Uncle Wal needed any parts or machines sent to him he would ring their daughter, Juanita and she would put it on the train for them to collect. This worked rather well. Wal and Anne had a van and caravan they travelled in and they camped at the back of the shop they leased for the duration of their visit. They would spend up to a month in the one town if there was a need, fixing and selling machines while there and from there they would place an ad in the local paper at the next town before moving on. When we were in the area they would visit us out on the stock route.

Aunt Anne took Mary with them for a week for a visit and the following week I went back with them and what a treat that was. Aunt Anne was and still is a loving, kind person and a delight to spend time with. Like Dad, she had a terrific sense of humour and while in Walgett a travelling show was on in town and Aunt Anne and I went. Half way through the show, Aunt decided she had had enough and asked me if I wanted to go and of course I said no – it was colourful, spectacular, funny and my very first outing without siblings and I wanted to stay. I saw my first and only ballet in Walgett. I never forgot that night and how much fun I had and I was ever grateful for my kind, loving Aunty who had enough love in her heart to bother with us. Sadly, I only had one visit, Mary had two.

Most drovers did not care too much about their sheep getting burrs in their wool but Dad did his best to keep his flock as free of them as he could. He had his very own "thrashing" machine, for bashing them down, his kids! Weeds like the Noogoora and Bathurst burrs get tangled in the animal's wool and are very hard to get out. The burrs contaminated the wool reducing the value, and increased the processing costs. The shearers hated to shear sheep that were covered in them. They were found in waterways and where sheep camped on a regular basis. Many a good campsite was ruined by a collection of these burrs and the council did their bit to control the pests.

The Noogoora burr could stand up to six feet high. If we came to a patch of these pesky burrs, it was us kids who had the job to get a stick or whatever we could get, and to knock down the burrs before the sheep went through. Most times we would "accidently" hit each other and have a good fight, laugh and run around. The job was not a lot of fun mid-summer as it was hard, hot and exhausting work in the unforgiving heat. We of course ended up being full of these burrs ourselves and then spent the rest of the day plucking them off each other. Our hair got full of them and really tangled.

If the patch of burrs was too large for us to knock down, we would bash down as much as we could in the time we had and then stay and help guide the stock around that patch. We would then run to the truck, have a drink out of the water bag hanging on the side and jump in the back, satisfied with a job well done and Mum would drive off to the next stop.

Any outing we went on was a treat and Dad found a way that he could trick us. After dinner one night, the dishes had been washed and we were sitting around chatting. Dad came out of the van and said, 'Righto you kids, have a wash, comb your hair and get dressed, we are going to the show!'

Wow, what a complete surprise. So there were seven little Australians all rushing around, eyes big, chatting with excitement as we washed and combed our hair and "smartened ourselves up a bit". Eventually we were all ready and stood in line and waited expectantly.

We younger ones were squirming with excitement and Dad stood in front of us, gave the seven of us a full inspection and said, 'It's the blanket show, now get to bloody bed.'

With tears of humiliation and disappointment, we trudged off to bed, I cried for ages with disappointment.

Dad pulled the trick again a few months later and all us younger ones fell for it again. We were rushing around getting ready and Emmie quietly said to us in passing, 'There is no show.' We still played the game and got ready. Deep in our hearts I think we hoped that this time it is not a joke. Eventually, Dad lost interest in this little game and left us in peace.

There was always great excitement when we really did go to the show or circus. New shoes, dresses and shorts were needed as these treats were so far apart, we would have all outgrown our "good" clothes. There was a week or so of preparing to get things in order with Mum and Emmie knitting new jumpers for us.

I recall an occasion when there was a drover a couple of days ahead of us. Mum and Beryl were in the front of the truck. Beryl was standing up beside her with her arm around Mum's shoulder and Dad rode up to the truck cabin and he and Mum were chatting through the truck window and he said to her, 'That useless bastard drove his sheep right through those burrs. There are some crook drovers around aren't there?'

Beryl piped up, and said, 'Yes Dad, and you're one of 'em.'

This caused great gales of laughter and that comment was repeated many times over the years on different occasions.

A good drover knows his mob, and stock tend to travel in the same position: leaders staying in the lead and the tailers staying in the tail. There was nearly always a fence on one side and if it was not scrubby and the stockmen could see the sheep, the stock weren't always counted. The mob would be strung out as much as possible so they could all get a feed as they walked along. On private land, the drover could only legally string out the stock three chains (440 metres) over the road. Col would be at the front of the mob to stop them from getting too far ahead, Les would ride on the side of the mob and Dad at the back to keep the strays together. Emmie quite often rode with Dad or took the other side if there was no fence. Each man had a whip and a dog, but if the dog was whip shy, then the whip would not be used.

Droving in the "long paddock" the grass could get a bit sparse and dry. When there was a shower of rain, the water ran off the side of the road and built up, so after a bit of rain the grass on the verge could be quite green and lush. The sheep in the front of the mob got excited and they ran along after the green pick. Col, the leader, had the job of keeping them under control. This was no mean feat as the woolly mob ran "baarrring" to let their mates know there was feed ahead. With a good dog this was easier but the sheep pushed all the way. With the noise they made, the rest of the mob wanted to walk on the road also, but if they did not get a taste of the green, they were content to stroll along getting a pick of grass from among the weeds growing on the stock route. Col was a dreamer and it was a lonely job at the front of the mob all the time. If he was not alert, the sheep would overtake him and then he would realise he was a way back in the mob. This created no damage, except for his pride. As he led the mob along the table drain he quite often found money, books and newspapers lying on the side of the road which he brought home to share.

It was a hot day and the night promised to be hot also. The sheep just trudged along from tree to tree and the men and dogs were working overtime to get them to the night camp. Because it was so hot the sheep kept wanting to run, thinking the summer haze was water in the distance. It got to be a difficult job to keep them within the mob. A mob this size had to be broken into three lots to let them fit comfortably around the smaller trough. As they all had a drink they tried to lay down to rest but we had to keep them moving to let the next lot come through for a drink. After the whole mob had been watered, they could wander back for another drink and laze around. Thankfully the water was where we set up camp so they could either drink, graze some more or have a rest. Most liked to graze if there were pickings of grass near the dam but this depended how much stock had been through before us.

Dad was late putting the sheep into the sheep break one night and tempers were ripe and the dust thick. Dad had a half grown dog that was not fully trained and as he tried to send the dog around the sheep, it got flustered and ran back to him again and again. Eventually, Dad picked up the young dog, swung it around by the tail and flung it into the mob of sheep. Luckily it landed on a fat woolly sheep's back, sat stunned for a few seconds, got orientated and finished its job of yarding the sheep. This dog eventually made a great sheep dog and it worked beautifully for any one of us.

The air was still and in the distance we could see the black clouds of an impending storm. Soon lightning was flashing and hitting the ground with a great show of power lighting up the whole area around us. We could hear the wind rushing towards us in the trees and we had to run around securing any loose gear that could blow around and frighten the sheep and make them rush. We quickly grabbed the table, chairs, dishes and the empty kerosene tin we used as a step and put all the gear in the back of the truck. Water came tumbling out of the sky with small hail making a racket as it hit the roof, thankfully the storm passed by with no mishap.

Some hours later another storm hit. The truck and caravan were shaking with gale-force winds. All us kids were cowering in bed at the time, scared of the loud bangs of thunder and the ferocity of the lightning and we expected the truck to roll over at any moment. Suddenly the truck motor started. Dad had gotten out of bed to turn the truck and caravan into the wind and just as he crawled back into bed he heard the noise of the sheep as they rushed. Bellowing in fear, they charged through the sheep break, galloping into the scrub in a frightened frenzy.

Dad yelled for us all to jump out of bed and help round them up. The tail end were still nearby; we could see them in the many lightning flashes. We ran like mad after the sheep and Col rushed to get the horses so the men could ride after the stock. Mary, Mike and I were all on foot, the wind was still shrieking and we could see and hear tree tops falling down around us. The rain was pouring down and we had no coats, shoes or hats so we were soon soaking wet. After the initial scare, we got our second breath and started to enjoy the adventure. We rounded up any sheep we could see and brought them back to the sheep break. Mum had in the meantime found the hammer and was fixing the damaged sheep break. We had to pull out the bent iron pegs and replace them with new ones. Mary and I were heaving together as we pulled three dead sheep out that had been smothered in the rush and there were two others with broken legs in the far end of the break. By this time there was only the tail end of the storm left.

After we finished, we went back to the camp and put the billy on in the van so we could get a bit of warmth into us with a hot Milo. We put water in a bucket and washed our muddy feet, towelled ourselves dry and changed out of our soaking wet clothes. We all had cuts and grazes from our tumbles and Mum put some Dettol in a bowl and we dabbed the brew onto our cuts. After we made our mug of Milo we excitedly spoke of our adventures: Mary was watching the sheep she was chasing and not where she was going and ran into a huge bush. She had to crawl out backwards it was so dense. She then took off running, chasing another sheep and they dodged each other around a tree. Mike had run into a tree and had a gash in his forehead. After shaking his head like a dog he then promptly ran into a low hanging branch and landed flat on his back. I had run full pelt into the fence chasing the sheep in the next paddock. I quickly got my bearings and took after the sheep on my side of the fence. Racing flat out I tripped over a log and landed face first in mud and water. What a sight we all were! The men came back with the stock and these were then bedded down for the night. The riders had a cuppa to warm up and we all went back to bed, although it took us some time to get to sleep as we kept reminiscing over our mishaps and exaggerating what had happened to us. Emmie got a bit fed up with the chatter and giggling and demanded we, 'Shut up and go to sleep!' What a spoilsport!

At daylight the men were back on their horses and the sheep

were let out of the break for a feed. Emmie, Mary and Les were watching them while Dad and Col rode out looking for any strays they'd missed the night before. Before leaving, Dad told me to skin one of the dead sheep for the dogs and give them a feed. I did not like skinning the sheep as it was stiff and awkward and heavy. My hands were sore from the cuts and bruises I had got from the night's chase. However, I managed to do it, while leaving a bit of wool on the carcass. The dogs never seemed to mind if the sheep was not dressed properly. I also disliked cutting the gut to let the innards out as it had a particular smell but a girl had to pull her weight and if that was the job – well you just did it. We never thought to moan and groan over these jobs as it was a normal part of life. After I finished skinning the sheep, I cut it up the best I could and Mike and I fed the dogs. The rest of the carcass we placed in a large bucket and placed a bag over it and then put it in the back of the truck to keep the flies and ants away.

The men found a few of the sheep and boxed them up. Col held them in one spot as Dad did a wide berth of the area, looking for muddy sheep footprints. He was satisfied they had found them all and brought them back to the rest of the mob that were lazily grazing around the campsite. As the sheep bellowed seeing their lost mates joining them, the men heard "baa-ing" behind them and they looked over their shoulders to see about fifty trotting to join the mob. They shook their heads and wondered, 'How did we miss that many?' and had a companionable laugh together.

We had lunch and Dad decided we would spend the rest of the day camped here. He skinned one of the other dead sheep to take with us for the dogs and dragged the third sheep away from the camp. We covered the leftover innards with soil to stop the flies and ants from congregating too close to camp, thus helping to keep the smell at bay.

The following morning Dad decided he had better tally the sheep out of the break. For some reason Mum could not come and help with the count, which was usually her job, and I got the unenviable job of being the count taker. I knew Dad was in a foul mood and I had never done this job before. As the sheep galloped out the small opening, Dad counted them. At each hundred he would call, 'undred' or 'Heh.'

I had to yell back, 'Yes.' This meant I had heard he had counted 100 sheep. A piece of paper and pencil was at hand and I wrote down 1 for each 100 that came through the makeshift gate. This day the sheep were frisky and as they streamed through the gate, Dad counted them using his fingers like a counter: if four sheep went through the narrow opening, he would have four fingers pointing at them or how many were trotting through the small opening. He found this method very successful. If too many tried to squash through the gate at once he would walk closer to the opening and the sheep would balk down to one or two. As they jumped through the small opening they would kick their heels up in the air or just gallop through the opening. On this particular day, as they came through the gate they spread out a bit on the other side. The uncounted sheep still behind the fence decided to take a shortcut to their mates on the other side by jumping over the fence. I noticed this and had to run back to try and count the ones that had jumped over the fence and stop the others from following.

This became very awkward as I had to stay within hearing range of Dad. Quite a few sheep became brave and made the leap to freedom so I had my work cut out trying to keep count on two fronts and stop the others from jumping over. Eventually Les, who was moving them up realised I was in trouble and managed to get his dog around to stop them. After all the sheep were counted we were 100 sheep short! Now there was a dilemma. Were we short 100 sheep or had I missed one of the "undred" grunts? Much to everyone's horror they had to be counted again – the whole 7,000 of them! After the second count the number was correct so I had missed one of the almighty grunts whilst running back to stop them from jumping over the fence behind me. Normally if Dad did not hear 100 called back to him, he would stop the counting up to that point and confirm that it was noted. Other times when he could not get someone to be the confirmer of the count, he would have a box of matches and he would put one match per hundred in his hand and this way he kept track of them. Needless to say, I did not get that job again.

Living as we did we often got a bit of bark knocked off our arms or legs but that didn't create any dramas unless a fly wanted to land on it. We would then wrap a piece of rag over the wound or get a bandaid – if we had any. The flies usually didn't bother us too much as the section where the break was always smelt of urine, sheep dung and wet wool and that was far more attractive than a bit of blood on someone's arm.

One day Mike was walking close behind me with a handful of steel pegs from the sheep break. He misjudged the distance as he went passed me and a piece of wire hanging off one of the pegs snagged the back of my shorts. At smoko Mum told me I had a tear in the arse of my pants. I looked over my shoulder and said to her, 'That's okay, it helps to keep the flies off me face.'

We were between Moree and Collarenebri and came across a truck that had rolled. It carried cartons of Coca Cola and bottles were scattered around. Very few were broken which showed the quality of the glass. We stopped and had dinner where the driver was camped. He had very little food and comforts and was waiting for another truck to come and pick up the load. He very generously let us have a carton or so, which we thought was wonderful, a rare treat indeed. He had dinner with us and the parents left him some food to see him through until the other driver arrived.

Mike was prone to getting hurt or hurting someone else. When we all had dinner in the caravan, with three to four each side at the dining table, it was always a tight squeeze. Often adults and kids gnawed on a bone and when finished we just threw it out the door or window if it was open. Mike was the worst as his aim was never true and he would often hit the biggest target, which was Dad. When he hit us, we always got offended thinking it was aimed at us, but he said he thought he would miss everyone, so of course it became the ducking game. If we were sitting around a fire and he threw a bone, it would still hit someone and if we had a mug of tea beside us, it would be Mike who would kick it over as he walked past. As soon as you saw any movement, you automatically picked up the mug that was on the ground beside you, no matter who moved, to save your cuppa.

After a while it became a natural movement. If you put your meal on the ground while you got a slice of bread or more sauce, someone would walk past and kick dirt into the meal, so you soon learnt to make sure you had your meal all prepared before you sat down. Dad had it all under control, as he just got one of us camp kids to get whatever he needed. Another thing to trip over was the three-legged tripod used for cooking over the fire. It was awful if you tripped and even worse if there was a pot on it, because it spilled everywhere.

In the early days we only had one kerosene lantern. We had to cook, eat and wash up by that one light. Eventually, Dad bought a gas Tilly lamp. Once lit, it had to be pumped on a regular basis to keep the glow high enough to read, play cards or do the dishes. It had to be handled very carefully as you could easily break the mantle inside. We always had a box full of spares. The glass never got cleaned until a mantle had broken, which was a regular occurrence even though it was carried in a bucket with an old towel wrapped around it as a buffer to rough treatment.

When we were at any campsite for a few days, it would get untidy and Mum would tell us to, 'Clean this mess up, it looks like a black's camp.' Cleaning up the camp meant to put anything not needed up against a tree trunk or the tyres of the truck, or placing it back in the truck. There were never any empty tins or papers lying around as paper was thrown into the fire and tins were buried outside the camp area. The men had a tendency to put their work gear wherever they felt like. If there was not enough shade under a tree or we were not near any, we would all sit on the shady side of the truck or caravan for relief from the heat. If no shade we then sat in the truck or caravan and sweated it out.

The sheep were looking a bit poor so Dad decided they had to be drenched. He organised to use a cockies' stock yards and he set to and drenched the whole lot himself. It was a mammoth task with over 7,000 to be done. We were camped a short distance away and the news was on the radio. Mum told me to run and tell Dad that the President was dead. It was 22 November 1963, John F Kennedy had been assassinated in Dallas, Texas.

Dad shook his head and said, 'The good ones don't last long mate.'

A bit after this, Dad had to re-shoe one of the mares, Penelope. He led her into the shade of a large box tree. She was so quiet Dad just left the reins hanging on the ground. He told me to light a fire nearby which I did and I gathered some bigger sticks and small logs to keep the heat and coals going. He gathered up his shoeing gear. First came the anvil, a hammer, rasp, hoof clencher, hoof cutter, pinchers, large pliers, horseshoe nails that were kept in an old Sunshine milk tin and a potato bag with horse shoes of various sizes. He gathered a handful of likely fitting shoes to try on her. He threw four shoes into the fire and took up one of her front feet, held it between his knees with his back to her head and used the hoof cutter to trim the hoof. Once he got it near where he thought would suit the shoe, he got a red hot shoe out of the fire with the pliers, put it on the anvil, made corrections, put it back into the fire, let it get red hot again and repeated the procedure until he thought it would fit.

Once again he picked up her front hoof and stuck it on, smoke smouldered and the mare tried to jump away.

'Here Chad,' he yelled, 'come and hold 'er head still for me will ya.'

I trotted over, my skirt twirling around my calves and as I got to the mare's head he said firmly, 'Now hold her head still.'

I nodded and picked up the trailing reins and held them firmly just under her mouth.

Dad picked up the shoe again out of the fire, grabbed up her foot and put the red hot shoe on her hoof. She pulled her head up and started back in a jerk. Dad snapped, 'Hold her f...ing head still will ya.'

I stood straight in front of Penelope, grabbed hold of a chin strap in each hand and looked up into her eyes. Here we stood, looking at each other. Dad once again grabbed the red hot horseshoe with the pliers and plonked in on her hoof. She jerked her head and I was lifted up in the air with her, a foot or so off the ground. Penelope put her head down and I was standing on terra firma again. Then she lifted her head again, taking me with her again. In the background I heard laughter and knew the family had noticed me swinging. Penelope gave a shake of her head and I of course was tossed around. With me still up in the air hanging on with the cheek straps she turned her head to one side and she then turned her head to the other side. I bumped lightly into Dad and he was not amused as I must have pushed him off balance. Penelope straightened up and lowered her head and I was back on the ground again. My weight must have taken its toll on her! Penelope was finished performing and she just lowered her head into my chest and let Dad do what he had to do. I did not appreciate her blowing hot air out of her nose, spattering moisture all over my face and front – one way of getting her own back I suppose.

Horse shoeing is backbreaking work and if lucky Dad only had to do one or two at a time. All the riding horses were shod at the start of a job if the land was stony and they needed shoeing. Dad got a bit annoyed when one lost a shoe as the job had to be done again, but that was droving for you when on rocky ground. If a shoe had fallen off, he would do both matching feet to keep them even. If I had to hold one of the really cranky horses, it was not fun. They could and would stand on your feet as they moved around. If they were bad enough, Dad would tie their back leg up in some type of hitch so they couldn't move but it was awkward as they could lose their balance as the other foot was lifted.

Later on in the day, Dad brought another horse to be shod and this one was not as quiet as Penelope. Even before he picked up a hoof, the horse started to fuss a bit, shaking his head as if to say, 'No, no.' The deed had to be done and I was once again the mug who got the rotten job! With my previous experience I tried to hold onto the cheek straps again, but he just reared up and went backwards and I quickly let go of the straps. Dad hit him on the head with a whip handle but that did not settle him much. I tried talking to him and he settled a bit until Dad picked up his back foot and then he tried to get away. I held on as best I could but it was difficult. He reared back over Dad, knocking my head back in a jerk. I pulled him with the reigns as I fell over. No harm was done and eventually he was resigned to the fact he was getting shod. I ended up with a few bruises and learnt a few more swear words that I was never allowed to say – out loud anyway! I was probably more amicable to deal with and easier to control, so I guess this may have been why I got the awful dirty jobs.

We were between Collarenebri and Moree when Dad decided to buy a new truck. He left us camped on a reserve and he and Emmie drove to Sydney looking for a suitable vehicle. They stayed with Dad's nephew Ron Kemp and his wife Mary and their seven girls in St. Marys. Ron took Dad around Sydney looking at trucks and he chose a green International AS-160 four-wheel drive. This was the first new truck he had bought. While in Sydney, Ron and his wife took Dad and Emmie to the workers' club and saw the local hit, Wayne Newton. We heard about this trip for ages afterwards as Emmie loved Wayne Newton. We were sick of hearing her sing his songs. On the way back from Sydney, driving along the highway, pleased as punch with his new vehicle, the TRB (Transport Regulation Board) pulled Dad up and asked him for his log book.

Dad said, 'What do I want to carry a log book for? I don't carry logs.' At this stage he had not heard of the larger vehicles having to log their trips in a book, so another lesson learnt although they let him off with just a warning, no doubt having a laugh at his expense.

We had the truck for a short while and Dad was longing to show off. It had been raining for ages and there was water everywhere. The wire and star pickets for the sheep break had to be taken to the fence line, which was a fair way off the road, to make the sheep break for the night. As it was too far to carry all the stuff in the rain and mud, Dad decided to take it all over in the truck. He lined us all up to show us just how good the new truck was. The table drain was deep with water and not far into the black soil the truck got bogged down to the axle! Dad was disappointed as he was under the impression that a four wheel would not bog. He was embarrassed when he had to get one of the nearby station owners to pull the truck out of the bog the following day. The tractor had trouble pulling it out of the wet, gooey mud as it was so deeply embedded. Interestingly, the driver did not take the tractor off the road to do the deed as he knew he would bog that also. It became a huge joke for us kids as Dad rarely came unstuck with anything, not that he admitted to it if he did. He had an audience that day so it was hard to live down the embarrassment.

One time when we were camped on a reserve near Moree, Dad decided that Emmie should accompany him on a trip to do the shopping. He wanted to teach her how to drive. Emmie drove to Moree and they swapped places before getting into town. After the shopping was finished and they were out of town, Emmie took over again. She was very nervous as Dad was not a patient teacher. All was going well and the truck drew close to the camp, which was on the left, off the road. Dad said, 'Turn here, mate,' and without slowing down and showing the instant obedience he usually demanded, Emmie turned off the high-built road and into the bush and hit a small gum tree. We were all watching, keen to see the new driver doing her thing and we saw her smash in the right side of the truck. Surprisingly, Dad did not kick up much of a fuss about the incident. Emmie would have been very stressed and I was able to sympathise with her 4 years later when Dad taught me to drive. He was very impatient. Emmie said she was so distressed, she could not look at the damage done to the truck. The rest of us kids thought it a great joke, the second favourite daughter smashing in the brand new idolised truck. We of course had to hide our mirth as we would have been given a darn good hiding.

We met a truck driver from Flinton Queensland who passed us on the road quite often and always stopped for a chat. His name was Mick O'Toole and he used to be a drover before the guts fell out of the business and he decided to buy a stock truck to join the livestock trucking business. A classic case of: if you can't beat em, join em. He also did the occasional small droving trip, but many a time Mick O'Toole was drunk while driving and he would stop the truck and fall out of it instead of stepping out. We often wondered how he managed to drive without running off the road and killing himself, someone else or the livestock he was carrying. His language was so bad that when he called into the camp Mum would send us kids out of hearing range. I remember him once saying how a mutual friend wore her shorts so short he could see her pubic hairs hanging out. He had a son called Tony, who was around six years old and quite often travelled with his dad in the truck. Many years later, Tony met and married one of Mum's niece's daughters in Toowoomba – Trish King. I have always wondered why we could not listen to Mick O'Toole's bad language when Mum and Dad used swear words on a daily basis. They were used as often as we would say "please" and "thank you". Dad had the knack of not swearing if any visiting females were in the camp. Instead he used the word "Blinky".

We were droving through Garah when Dad met up with Bud Williams, another drover who had a large bogey wheeled caravan for sale, which Dad bought. We kept the "Caravel" until we finished this trip as the larger caravan needed a bit of work on it. Bud let Dad leave it in his back yard till we came back to claim it.

We finished this trip in Meandarra as planned with the sale of the stock very successful. They were all bought by the locals, which excited Bruce Knobbs as his reputation grew even bigger. While there the decision was made by some local graziers to get together and buy a larger mob of sheep. Our next trip would be to drive their stock back into Queensland straight to their stations, not to the sale yards. So we had a job to start in the New Year.

MOREE 1963

Traditional land of the Kamilaroi people

We packed up and drove back to Moree, camping a couple of miles out of town on the Collarenebri road, on a reserve called Brewery Flats. It was the local "Lovers' Lane" and the reserve was full of large gum trees, briar bushes (Boxthorn), tall grass and plenty of meandering tracks leading off the road. The reserve was full of bird life: galahs, magpies, crows, finches, kookaburras, cockatoos, parrots and great flocks of budgerigars. The budgies were a real sight and I had never seen so many before or since that time. There were so many that they cast a shadow as they flew overhead, squawking and flashing their pretty green colours.

After we settled into the camp, Dad drove to Garah to pick up the new caravan and bring it back to Moree. It was much bigger than the old van and it had a small room over the front wheels, which was to be Beryl's bedroom.

She was at long last moving out of the parents' bed. In that tiny room was a bed with some storage space underneath and a cupboard along the opposite side that backed up to the table in the next room.

The caravan needed quite a lot of work to be done. Several windows were broken, the ceiling leaked and the front of the van had disintegrated and needed to be replaced. Dad decided to fix it all himself. He bought a full length piece of masonite and to make it bendable he placed it flat on the ground on top of a tarp, covered it all over with potato bags and wet it. It became our job to cart the water to keep it all wet. The idea behind this was that wetting the masonite would make it become flexible so it would curve around the front of the van. A few days later Dad pinned it on and used aluminium piping around the edges to hold it in place. After completing the job, he stood back to admire his handyman work only to discover he had put in on back to front! The ridge side was in full view outside and the smooth side was on the inside of the van. Dad made it a bit of fun saying how good the inside would look and decided it was to stay there as it was.

The top of the caravan was not waterproof, so a green tarp was bought and cut to size and that was glued on. Dad then painted the bottom of the van green, and the top red. Upon measuring the windows to replace the broken ones, he then discovered that nearly all of them were crooked and out of alignment. The glazier questioned Dad about his measuring ability but he still managed to cut the glass accordingly so we had new glass windows. The fly gauze also needed replacing. Unlike the other caravan, this one did not have a fly-proof door but we happily sacrificed that for the extra space it provided.

All the electrics also had to be replaced but thankfully it had a gas stove and fridge that worked. At long last … real butter! And meat could be stored for longer periods, we felt like kings.

The fridge and stove were on one wall along with a wardrobe and on the opposite side was a set of cupboards and drawers. Above the drawers was a dressing table with a mirror. This mirror was covered religiously if a storm was brewing as Mum had a fear of lightning striking the mirror. At the back end of the caravan was Dad and Mum's double bed. Above the bed were more cupboards that we filled with clothing.

Under the bed was a large storage area. It had small double doors that opened to a black pit. We stored things in there that were not needed daily. Mum's sewing machine lived under there along with a travel bag and when something was needed, Mike or I had to wriggle in to get it. Sometimes we had to throw bags and boxes out to make room for us to get in to rummage for what was required. We kept our school ports in here at the front in case we had a chance to get some work done. It was time consuming to set the work out and then settle down to do it, so we would try and learn our times tables by rote or read our readers. Only Emmie was a scholar. We others did not care too much about learning.

A favourite prank we played was to lock the door while someone was in the cupboard. I hated that dark space and would get in and out as soon as possible. If someone did lock the door, I would scream and yell and push on the door until it was opened.

Many treasures were found under there, as Mum used it as a hiding place and things were then forgotten about, such as tins of lollies. This space was also very dusty so on occasion we had to sweep it out. That was a huge job because everything had to be taken out first. Jobs like this were done at dinner camp while waiting for the stock and men to catch up.

The small room designated for Beryl became mine as she refused to sleep on her own. Some of the daily camp gear was placed in this room for travel from camp to camp: buckets, the enamel dish, towels etc. At night camp all these things were then placed outside to be used. I loved my little room and on the bright moonlit nights I could read a book with the bright light coming through the window. This also became my "resting" place whenever the truck stopped. I thought I had it made! This left Emmie and Mary sleeping in the back of the truck and the very first winter I made myself a wagga rug (made out of potato bags covered in one of Mum's old dresses). I missed the warmth of my sisters but not the waking up looking for the missed blankets that they'd managed to tug off me. We had this caravan until we finished droving in the late 1960s.

We settled in Moree to stay over Christmas. The horses were in another reserve that was fully fenced on the other side of the river. On occasion the boys would bring some horses to the camp to ride and we who could not ride bareback were encouraged to learn. It took a while for me to pick it up as I just could not get my balance, though I rode the saddled horses without stirrups. I had many falls while learning this bareback trick. It gave everyone a good laugh, and of course that pleased me no end. After all, it was Christmas! The trotting was hard to grasp as the horse seemed to be so rough. When I was comfortable with that, I had to master cantering. I found this a lot easier as it was a smoother ride and I flowed along with the horse. Dad was impressed with my "seat" and told me to canter out for a bit and then come back so he could watch me. I left the camp and gave the pony a good run and as I did a sharp turn around, I nearly slid off his back. I managed to regain my balance and I cantered at a good pace back to the camp with a big grin of pleasure on my face. It was so rare to get any praise from Dad. As I passed the briar bushes closer to the camp, Les ran out at me waving a bag around his head and shouting. The horse shied and I fell off with a thump onto the hard black soil still holding the reins. I stayed on the ground for a few seconds, getting my breath back. Les was bent over with laughter. After realising I had not broken any bones, I stood up, put my head down and raced into his midriff full pelt.

'You rotten mongrel bastard,' I yelled.

'Patsaay,' Mum shouted, 'I'll wash your f...ing mouth out with soap if I ever hear you use that language again.'

Dad still enjoyed a drink at these times and as the jobs were still found mostly in the pubs, that is where he went. Drovers or stockmen could either stand outside the pub, leaning against the wall and chatting to others or choose to go inside and have a drink. Dad nearly always chose the latter! He would drive into town with the idea of meeting his mates and enjoying the local gossip and come home drunk. Mum hated him being away from her. She needed him by her side at all times as she had a jealous controlling streak. One day Dad was dressed up in his fawn slacks, white shirt and freshly polished boots and when he tried to start up the truck it would not work. He peered under the bonnet and discovered the distributor was missing. Mum had thought she would outsmart him and hide it. As he was getting dressed, she had walked around to the truck, lifted the bonnet, grabbed the first thing she had seen and hidden it. Dad called out in a concerned voice, as if something had seriously gone wrong, 'Beryl, have you taken the distributor out of the truck? Bring it back or the truck will blow up quick, bring it now.' His voice was high pitched as if it was really urgent. Mum, of course, fell for the trick. She raced to Dad with the distributor and then hunted us all out of the "explosive" zone so Dad could safely replace the distributor and save his family from harm.

Dad got back into the truck, looking at Mum, and shook his head as if to say, 'Saved you all, just in time.' He then said, 'I will have to start the truck love, to make sure no damage is done.' The truck started first go. He put it into gear and drove off laughing. Mum immediately woke up to the trick and as he drove off she raced after the truck. Eventually she stopped out of breath and bent over, calling him every foul word she could think of. She trudged dejectedly back into the camp, smalls puffs of dust drifting around her feet. She was cursing him for the rest of the day. We all thought it was a huge joke but had to hide our mirth or she would tan our backsides. We spent a happy day having foot races, playing hopscotch and quoits and we stayed away from Mum as much as possible.

Dad did not come back until very late in the evening – very drunk. By that time, Mum was twice as angry as when he left. Being a happy drunk, he thought it very funny. Mum was furious to see him so happy and falling over drunk, she threw a bag of potatoes at him and missed with every one of them. She got angrier and angrier as he ducked and weaved around the hurling spuds. Even now, fifty years later, I can still see the spuds flying all over the place – it was hilarious. Our job was to pick the spuds up but that became another game for us too as we re-enacted the play! Our aim was not much better than Mum's.

Next day, the potato bag was leaning up against the tree trunk and a mouse sneaked out of it and started to climb up the tree. Someone yelled, 'Mouse,' and I raced up and grabbed it. At the same time, Mary grabbed the axe that was leaning against the tree and with the back of the axe hit the back of my hand. If it had been with the front of the axe I would have lost my left hand! However, nothing was broken, it was just very sore, with skin knocked off, and swollen for a while. Mum bandaged it up, gave me a couple of Veganin pain killers. A near miss for me and I think the mouse got away safely too.

Soon after this, Mike saw a lump on the side of the tree and it was moving very slowly. He poked it gently and it exploded into hundreds of spiderlings that rushed everywhere. It was the greatest phenomenon I had ever seen in nature although I don't think the mother spider was amused.

After a visit to town Dad brought home a set of Russian Matryoshka nesting dolls for Beryl. They were wooden, very colourful identical dolls that fitted inside each other, about five to the set. I had never seen anything like them before and I was very envious. On another occasion, Dad won a small mohair knee rug, which Beryl also received. I think he may have had a guilty conscience as next time he bought Mum a beautiful Chinese coffee set. It had the coffee pot, milk jug, demitasse cups with saucers and matching bread and butter plates. It was a set to behold with a raised dragon on each piece and it was lavishly set with gold edging. Sadly, it was never used and over time it all got broken. It was hard to protect something so delicate, living rough as we did.

We had our camp fire under a large tree and one night I saw movement out of the corner of my eye. I looked up and saw a creature I had never see before and blurted out, 'Look at the cat with a baby on its back.' Mum and the rest of the kids came running over for a better view and Mum said, 'That's a possum.' This gives you an idea of how little local fauna we saw due to the noise we created around us. Mum told us a story about how, when she was young, there was a possum who would climb down the veranda post on a nightly basis and our grandmother would let Mum feed it bread and jam.

Because we weren't on the move, we were able to save our vegetable peelings, meat and other food we did not want and make it all up into a stew for the dogs. A few handfuls of dog biscuits were thrown in to make it tasty and the dogs seemed to really enjoy it. Certainly, it always smelt good but to my knowledge, none of us kids were ever fed this stew.

Even though we were camped in the one spot, there was no water close by so it still had to be carried in the truck. And of course, as when droving, we still only bathed when "necessary" and we would have a wash in between.

Mum had a saying, 'Wash down as far as possible and up as far as possible and leave possible for tomorrow.'

This meant face, arms, neck and legs up as far as our shorts. We ran around shoeless so our feet were always filthy and not always washed before bed time and apart from the parents, we never used sheets. I shudder to think what the beds must have smelt and looked like!

Eventually, the boys discovered the hidden spots of our Lovers' Lane campsite, where young couples did their courting in so-called seclusion. The boys snuck up to cars and peeped in. If the couple were "doing it" the boys would tap on the window and then race off. Sometimes we younger ones joined in. At night, we took a torch and shone it in the car and then ran away, screaming with laughter. We were chased a few times but never caught as we had our escape all worked out and the older ones would grab our hands and tug us along with them, flying more than running. If our parents had caught us doing this, all hell would have broken out, but we got away with it. We were careful to keep this secret from Emmie. Les wouldn't tell as he was the first one to start it. I actually did a bit of courting there myself some twelve years later and I don't think anyone ever peeked at me.

Settled down for weeks like this were the times that we played "roundies" (bushman's cricket). We used a tennis ball and Dad would flog it into the bushes and get his six then stand back while we all had to search for the ball. We never won a game against him. An old kerosene tin was used as the wicket and of course we never hit it. Eventually, Dad would relent and then he became the bowler. He would always be far too fast for us and got us all out very easily. Dad loved to do silly things to make us laugh.

For entertainment we also read funny comics and played cards. Dad enjoyed playing euchre and as he openly cheated, keeping back cards that should have been played he always won this too. If he was caught out, he would get really indignant and deny he had done anything wrong. Mum refused to play cards with him as he was a sore loser if, on the rare occasion, he lost. In the end, no one wanted to play with him so that was the end of the card playing.

BOURKE 1964

Traditional land of the Nyemba people

On these trips we nearly always went from Bourke, Walgett, Moree, looped back to Narrabri, Wee Waa, Coonamble, Pilliga, Collarenebri, Garah and then into Queensland's Goondiwindi and Meandarra.

Colin was now eighteen, Emmie seventeen, Mary sixteen, Les fifteen, Mike twelve, Beryl eight and I was fourteen. Colin had long ago left school while Emmie, Mary and Les left a few years later. We three younger ones still struggled with the basic three "Rs": readin', 'ritin' and 'rithmatic. Mike and I were in the same grade and we struggled with our schooling unlike Emmie, who had been dux every year. Perhaps if we had had more study time and a patient teacher we would have achieved more – but we could only work with what we had.

This was our third trip to Meandarra from Bourke. Along with Sid Derrick there were the Penfold brothers, Fraser Horn and Noel Murphy buying into the mob. All these men came from Moonie, just outside Meandarra, Queensland. Once again the stock was sold by the yard full and there were just over 8,000 sheep for sale. The group bought the sale yards out, leaving not a sheep for anyone else. Like the last mobs, they all shared the same brand and ear marks.

We were a few weeks out of Bourke near Collarenebri and camped off the road in a sandy area. Next morning, as Mum tried to drive off the truck got bogged and the more she revved the engine, the deeper into the sand it sank. We flagged down a passing motorist to send word to Dad. He came cantering back and Mary stood holding the horse's reins while he ranted and raved about the truck being bogged. He unhooked the caravan and we gathered bags and small saplings to go under the back wheels. Eventually the truck came free. Dad drove to firmer ground and retrieved a long rope from the back of the truck and hooked it onto the caravan. For some reason Mum held onto the rope and as the truck took up the slack, it twisted her over the rope and flung her quite a few feet away. She got to her feet, a little groggy, and gave Dad a few choice words. He managed to pull the caravan free of the sand and hooked it back on to the tow bar of the truck. He must have been very annoyed as he cantered back to the stock instead of letting one of us ride to meet up with them as he would usually have done. After this incident Mum had a dreadfully painful neck, her head ached for days and she was sore and bruised from the fall.

After the night meal we would have a yarn or read a book if there was plenty of fuel for the lights. We quite often read by the campfire into the evening if it was not too hot to sit there while Mum and Dad would go to bed in the caravan and listen to the radio. If we were lucky enough, the radio would be loud enough for all of us to hear. Much to Dad and Mum's consternation, Col would pull his bed up to the side of the van so he could listen to the radio. There was not a lot of privacy for drovers, no matter what you were doing. The days were all fairly routine, broken only by other drovers being shown through a property or someone passing by who'd call in for a chat. When drovers called in we would catch up on what the others were doing, so I guess it became a gossip session. We loved it.

When we were close to Moree, we received news that Dad's mother had had a stroke, so he and Emmie flew to Sydney to visit her. On this trip Emmie bought Dad a photo album for his thirty-fifth birthday and it is still in the family today. While in Sydney they stayed with Dad's nephew Ron and his wife and family. After that visit, Ron and his family became a part of our lives. We spent a Christmas with them in St. Mary's. Ron, Mary and their five youngest girls came for a visit out on the road and spent a few days camping with us. Most of these visits were spent on a reserve so we could have quality time with them. They loved the novelty of the open fire and camp-oven cooking and of course, we envied them their formal schooling, bedrooms and the love they shared.

We were droving the sheep between Moree and Narrabri along the Gwydir highway and the train line that ran beside it. If the stockmen were not alert, the trains could sneak up on them and the driver would blow his horn as soon as he realised stock were loose ahead. When the train came along with a rush and a puff and a clatter, puffing black smoke, the loose horses would stand looking at it and as it went rushing by, they would gallop madly away, excitedly kicking up their heels, bucking and rearing and having a good time. Les would race to his horse, jump onto the saddle and gallop along the side of the railway track, cracking his whip, with Snip his dog only feet in front of his galloping horse, barking madly. It was a real show to see the sheep bellowing in fright as they ran off the railway line, some stumbling and falling but managing to get up just in time to get away from the monster that was rapidly bearing down on them, and the horseman galloping to save the stock from certain death. The people in the train would be shouting and waving to Les as they passed by.

'A job well done, mate,' they would shout. Les adored this attention and once the train timetable was set, he would wait until the train's whistle was blown and away he would gallop putting on a show for the bored railway passengers. The train whistle was sounded when there was something ahead that needed to be warned it was approaching.

When travelling near Wee Waa, Col met up with a young man working on one of the stations nearby. "Speedy" came to the camp in a lovely two-seater red sports car. He was going to the movies in Narrabri and invited us to go with him. As he only had the sports car, Col, Emmie, Mary and Les got to go along with him. They had the pleasure of seeing *North to Alaska*. That movie was all they could talk about for weeks afterwards and Mary told me everything about it. She remembered the movie so well and talked about it so often that when I saw the same movie years later, I knew just about every scene from Mary's vivid description of it.

If the stock were easy to handle, Dad would send Emmie ahead to the camp. She would have a good wash in a small amount of water and change her clothing and this would shame Mary and me into doing the same.

We were still between Narrabri and Wee Waa when the sheep had to be shorn and it was Dad's job to look for a station that did not mind hiring out their shearing shed to Winchcombe Carsons for the shearing. Eventually he found a place and we camped in the front of the sheep station and the stock was brought in daily in mobs to be shorn. Sheep cannot be damp when shorn so they were brought in late in the afternoon and penned under the shed and in the outside yards ready for the next morning. Col and Les held the sheep out in the stock route. As they were shorn, they were brought back out and Emmie and Mary would hold them at the other end of the stock route, far enough away so they did not get mixed up.

Fraser Horn worked for Winchcombe Carsons in Meandarra, so he came and helped with the shearing. Fraser had his dinner with us and Emmie made it her business to have her lunch at the same time. She had fallen madly in love with this older, refined man and spoke about him all the time – she was really smitten. Mike and I relieved the workers for their breaks and Dad and Fraser were the shearing shed hands. They yarded the stock and then brought the shorn sheep back out to the stock route.

The weather was still warm at this stage and we were there for just under two weeks. Mike and I really enjoyed spending time with the stock, pretending to be real stockmen. I had to stop trying to crack the whip as the horse did not enjoy getting hit and I quite often wrapped it around myself. I did okay when not on the horse but being that much higher up it beat me; I could not grasp the fundamentals of it. The rest of the siblings were quite good with whip cracking.

Mum kept complaining of a sore head, back and hips from when the truck was bogged and she appeared to get worse as the days went by. However, she didn't go to the doctor.

'What do they know?' she would say whenever one was mentioned. She was a real hypochondriac – if one of us had a headache, she had a migraine. If our back hurt, hers would be killing her. She was very funny in a way and loved getting attention from Dad. She would be chasing us around and doing stupid things and the minute Dad was near the camp she would start to complain that her hip was aching or her leg hurt. All she needed was for him to say, 'Is it love?' and she was content with that.

It was around this time she started complaining to Dad that he never took her anywhere for an outing. He would say, 'Get dressed love and we will go to town.'

She would retort, 'I don't have anything decent to wear.'

'I will buy you a new dress love.'

'We can't afford it.'

The conversation would go on: 'I don't have any decent shoes, no stockings etc.'

Dad kept nagging at her to come with him and one afternoon she really lost her temper and shouted at him, 'I don't f...ing want to go anywhere.' Sadly, she just needed attention. I remember thinking, *I wish he would offer me the trip*, as we kids rarely had a browse in any stores even if we went into town for some reason.

It was a really wet rainy season. Mosquitoes and sand-flies bred like mad and drove everyone crazy with their biting and constant buzzing. For relief at camp we would gather horse and cow manure and place it on the fire and stand in the smoke.

'What is worse?' we wondered as we stood there, putting our heads out of the smoke to get a breath of fresh air. 'The bites and stings or the smoke and runny eyes?' A hard choice but we managed to get a bit of relief with the burning dung. Collecting manure certainly made a difference from collecting wood! We would carry a bucket to put the dung in and only carry a couple of cow patties at a time as they were awkward to handle. It was a juggling act to keep them from falling out of our arms. When standing in the smoke one of us would say, 'Smoke follows beauty,' then the one who spoke would get pushed out of the smoke so another one would stand in it! Often one of us would end up in the smouldering pile of dung, laughing at the antics and brushing at the burn we would get, but we were tough kids. The cooking fire was away from the dung fire as the cook needed to be able to see what they were doing. It was not unusual for the quieter horses to come and join us in the smoke. We had to be careful the horses' swishing tails did not catch us in the eye. If the surrounding grass was dry then holes would be put in the sides of a bucket and the dung fire lit in that so it did not cause a bush fire.

We were quite often infected from the mosquito and midge bites and a tin of goanna salve was always handy for us to put on our sores and itchy bites. It was very hard not to scratch ourselves from the bites of these pesky things and they also drove the horses mad. They would feed on the white of the animals' hide. Dad mixed up a brew of black shoe polish and Stockholm tar and he spread this on the white spots. It had to be done every two or three days. Even the wildest horse stood still as he rubbed it in. They knew a good thing when they were on to it. Some old bushies used Dettol mixed with castor oil. This was good for repelling the mites, midges and sand-flies and it worked well with both animals and humans.

A Hawker called Wally Cook travelled around New South Wales and Queensland with a truck full of all sorts of things. Shoes, clothing, bedding, cotton, material, camp ovens and anything a person could want out in the bush. Mr Cook came through and offered to show Mum what he had for sale. It was so damp and wet and miserable and we were complaining bitterly of the cold. Mum climbed up into the back of the salesman's covered truck and he showed her what he had.

Mum bought us each a pair of "blucher" boots. They were ugly and bulky to look at but we were very grateful to receive them. I remember mine did not fit too well so I put a big rag in the toes to take up some space, as it was that pair or nothing! We all had feet covered with blisters for a while but we persevered until the boots became bearable as we knew we would not get another pair until summer, if they were needed of course. Mum would say to us kids, 'Beggars can't be choosers' and we lived by that mantra in all aspects of our lives.

When Emmie became one of the "stockmen", she took Judy as her working dog. Dad did not mind this as he bought a dog in the mid-1950s called Speck who was a one-man dog and would not leave his side or his horse. If we rode Dad's horse, Speck would come with the horse, walking in its shadow but would not work for us at all. No amount of begging, crying or cussing would make that dog do anything and he would look at us as if to say, 'You are not my boss.' He would do anything for Dad as long as it was Dad doing the asking. He was so well trained that once he learnt where the barriers were, he did the job on his own. When in dinner camp and the sheep were not to pass the campsite, Dad only had to say, 'Get him Speck' and the dog would trot out and turn the sheep back into the mob to where they should be. The next time he would look at Dad, who would flick his finger and away Speck would run. After that he just did it without being asked. Dogs like Judy and Speck are very hard to find and they were the best I have ever seen. Dad would say he wished he could train his tribe of children to be so obedient!

Bruce Knobbs had asked Dad to keep an eye open for a good pony for his children and on our travels we bought a Shetland pony for them and we kids had the pleasure of helping to get the pony quiet. The only time I have ever been bucked off a horse was on that little black and white pony. That pony was as rough as guts to ride but I believe all Shetlands are like that. Once the pony was quiet enough to Dad's satisfaction, he encouraged Beryl to ride it along with him whenever she wanted to. Beryl thought this was great so she begged him to get her a Shetland pony.

One of the most horrifying things to happen to us was seeing Mum have her first convulsion in front of us when we were doing our schooling. Mary, Mike, Beryl and I were sitting in the front of the caravan at the table, two on each side, and one minute Mum was lashing us with the wooden ruler, trying to convince us that ten plus ten equalled twenty and the next minute she was on the floor, convulsing. We were all terrified as we had never seen anything like this before. Mary jumped over Mum writhing on the floor and ran screaming towards Col who was coming into view over the horizon leading the sheep.

Col heard the frantic screaming and saw Mary running towards him and galloped towards her, and after he got the message he galloped back to tell Dad, who was still way behind the sheep. Dad sent Col galloping off to the nearest homestead to call an ambulance from Narrabri and tell them that he would meet them halfway in the truck.

Dad assessed how Mum was and then carried her to the front of the truck still unconscious. He placed her on the seat and told Emmie to get into the truck and hold her up so she did not swallow her tongue and to keep her from falling off the seat. Dad quickly disappeared in a cloud of dust as he drove the truck towards town. This left Mike, a hysterical and distressed Beryl, and me in the caravan. It took a long time to settle Beryl down and we were all distressed, thinking we had lost our mother and would never see her again. I cleaned the caravan floor, put away all the school books and tried to distract Beryl who kept looking up the road expecting to see the truck. Mike and I entertained her as much as we could but she kept breaking down sobbing her heart out.

While they were away Col, Mary and Les did the droving. They slowly brought the sheep in to water at the trough and then let them spread out to feed. As the sun began to set, they started bringing them into the camp. At this stage the sheep break had not been put up but fortunately all the gear had been taken out of the truck. Just before sundown, Col rode into the camp and helped Mike and I put the sheep break up. Beryl dragged a few pegs and we jollied her along as it helped to keep her mind off what was happening to Mum.

I stepped up and started to get tea ready: roast leg of lamb, potatoes, pumpkin and a tin of peas followed by custard and tinned fruit. It felt like forever before the dogs started to bark and we knew the truck was pelting down the road towards us. We all expected the worst but Dad assured us that Mum was going to be okay. The first few visits, Dad and Beryl went on their own as they had to leave mid-afternoon to fit in with visiting hours. After a few days Dad decided that the rest of us kids could go and visit too.

We were quite a few miles out of Narrabri and Dad put the truck lights on as we were driving along. Mike asked why and Dad replied, 'It's getting a bit dark mate.'

Mike replied, 'No it isn't, it's still light.'

Dad realised he had his new sunglasses on and they were making it all very dark! It took a few years to drop that subject too as up until this time Dad thought only "mug lairs" wore sunglasses. Dad had started to get cataracts on his eyes and the dark glasses no doubt were a good thing as being constantly in the dust behind the sheep did not help his eyes.

Mum was in hospital for over a week and when she came home, she was a different person. She was dazed and did not remember anything leading up to the convulsion and often sat in long silences, staring at the wall. She was always tired and slept a lot. Until she felt able to drive again, Dad took on the driving with Mary or me riding his horse in turns. If it was scrubby and there was a possibility of us losing stock, we would canter his horse to where he'd parked the truck and he would ride back to the stock and bring them onto the truck. By this time it would be near smoko or dinner time. This became the routine and we did this until Dad had to take Mum to Sydney for more tests.

Leaving us all on a reserve with the stock, Dad, Mum and Beryl travelled to Sydney and she went to a large hospital. Dad and Beryl stayed with his nephew Ron, Mary and their girls for a few days and then, leaving Beryl with cousin Ron, Dad came back to the camp. Mum was in hospital for over a week where she had dye pumped into her head and other tests were made. She and Beryl came back by train a fortnight later. She still slept a lot and took fits regularly; thankfully for us kids it happened mostly at night. We were quite traumatized from seeing her taking a fit.

Life changed for us all after that. We were taught to put a round dolly clothes peg between Mum's teeth, so she did not bite her tongue, and let her have the fit and then comfort her and make her a cup of tea. She would then sleep for ages. She still kept driving for a few years after this, but thankfully she never had a fit behind the wheel. At times she would have a pile of about fifteen pills in front of her that she had to take at the one time.

Some weeks after Mum came back home with her medication, she and Dad had one of their common and frequent arguments. As the verbal fight went on, Mum made ludicrous accusations. Dad really lost his temper and told her she was mad. Mum promptly jumped up and went into the caravan, rummaged in the port kept for business papers and she brought out a certificate to prove she was not insane. Dad took the slip of paper off her and said, 'Now go and get the friggin' piece of paper to prove that the f...ing doctor was not mad.' Those who heard this thought it was a real joke and we all had a good belly laugh. Poor Mum, she could never take a trick from Dad.

Mum now found it difficult most days to put the sheep break up so Mary, Mike and I took the job on. The first few tries were a laugh and a half. Mary unrolled the wire with her feet, pushing it ahead of herself while Mike, Beryl and I brought along an arm full of iron pegs each. Mary unrolled the wire in a half-moon circle like we had seen Mum do for so many years. First we had to thread the peg through the wire and then put it against a wooden fence post and either hammer it in or tie it onto the post. Then we would walk to the opening at the other end to thread an iron peg through the wire and hammer it firmly into the ground. This was about a 15 feet gap for the sheep to come in. Mary could reach the top of the pegs with the hammer. We had trouble getting the wire firm enough to keep the sheep in but after a few mistakes we had it under control. Mary pushed the iron pegs from the inside of the break and I held onto it firmly from the outside as she pounded the iron peg into the ground until it was firm enough to hold the sheep. Mike and Beryl kept bringing the iron pegs to us so we got the job done eventually. Dad was happy with our first effort, though he walked around the sheep break giving each peg an extra hammering to make it quite firm. We were pleased we did such a good job as he was a hard man to impress.

I took over the bulk of the cooking. Mum would give me the menu most afternoons but on occasion I had to do it without her input as she would be asleep. Lamb roast quickly became my *forte*. Of course the base of the meal was always lamb and potatoes and what other fresh or tinned vegetables we had. Cauliflower was on the menu quiet often and Mum made the most delicious onion white sauce to go with it, which I learnt to make. So even though they had a new cook, the menu did not change. I found it a challenge to make a stew as the base was hard to do and like Mum, I relied heavily on packets of soup and Vegemite. In winter, I tried to make a hot pudding every night – syrup dumplings, bread and butter pudding, creamed rice or just hot custard served with tinned fruit.

Dad bartered for a Shetland pony for Beryl. Cheeky was a white and brown piebald mare and bigger than Bruce's pony but far smoother to ride. Beryl wanted a "proper" saddle to ride on so Dad took her shopping and she chose the most expensive on show. It was made of light brown pig skin and it came with a matching bridle and he also bought her a pair of jodhpurs and hat to complete the outfit. We had Dad leading Cheeky with Mini Me riding the stock route – they made a cute pair.

Cheeky was a different matter. At dinner time she got a hub cap half full of oats or chaff to nibble on. This particular day Dad wanted to be off and he told me to go and see if she was finished, as Cheeky ate with her head in or down near the hub cap. I walked to her and I learned over to look into the hub cap and quick as a flash, ears right back and bared teeth, she grabbed me by the back of the neck, shook me like a dog and flung me. Dad raced over to save me, waving his hat and shouting but the deed was done. He gave her a hard swift kick in the gut and she retaliated by kicking out at him but missed. This was the first time she had kicked, so Dad got quite a surprise and so did she when he hit her again. My neck was badly bruised and I had trouble turning my head for weeks afterwards.

Winter was closing in so the jumpers and cardigans were brought out. Dad always wore long-sleeved flannel shirts with the sleeves rolled up, a hand knitted V-necked jumper and under that a hand knitted V-necked vest. They were usually green or blue although later in life Mum began using brown. All Dad's hand knits had cable or rope stitch and for variety some were double cable. We kids had different patterns on our jumpers and cardigans. Mum was a terrific knitter and Emmie was quickly taught the skill as well. Neither of them ever sat without knitting. From the front of the truck you could hear the needles clacking away. In the summer time if it was not too hot, they would knit in the cool of the afternoon and they always had a tin of baby powder nearby to keep their hands dry.

On the really cold days or when it rained, Dad wore a long oilskin coat as did Col, Emmie and Les as they were out with the stock. These coats hung over the saddle and the horse's rump to help protect the saddle from the weather. If we had hired men, they had to supply their own.

It must have been a year for good will as Dad was still doing a bit of horse dealing and he bought a nice horse and gave it to me.

'Here Daught, I bought this one for you,' he said to me. As we girls were getting older, we had become "Daught" to the men of the family. I suppose Daught, short for Daughter, was easier than to try and remember our names.

It was not a big horse that Dad had bought me, greyish-brown coloured and we got on very well. I called him Gayboy after a horse in the Melbourne Cup. I used one of the leftover saddles that we still carried in the crate above the truck cabin and this meant I had stirrups I could reach to keep my legs at a comfortable length. I was getting the occasional full day's ride to relieve Dad at times as his dermatitis and back were both starting to bother him. He drove the truck on these days giving Mum a break. I loved my days as one of the workers, but they did not happen often enough for me as I was still needed as the cook.

Mum enjoyed romance magazines in the later years – *True Romance, True Story* or similar. They were pure drivel but made for light reading and the fun of immersing yourself in a pretend world. I must confess, even though Mary and I loved to read, we did not know all the words so we just took a guess and carried on reading. I do not remember if we had a dictionary to look up any unknown words but there were many I did not know. As I got older, I would hear a certain word and think to myself: *So that's how that word's pronounced.*

When I was in my mid-thirties and living in Dalby, a good friend's father was an avid reader and I quite often swapped books with him and his wife. One day I asked him to explain what *coup de grace* meant. Of course I asked, 'What does 'Coop dee grace' mean?'

'Huh?' he said.

I showed him on the page and bless him, trying very hard not to laugh, he explained "coo de gra" to me.

Another of life's little mysteries solved.

Most nights after dinner we all went to bed early as we were up at daylight or before. Some nights we would play 'hidey go seek' in the dark. The sheep never seemed to mind the rowdiness, though the dogs would bark madly at us.

The Moree Annual Show was advertised and the parents decided we would all go as a family. Mum regularly received a catalogue for clothing and other goods and for the show Mum decided to buy all our clothes out of the latest one. We were measured, colours and sizes agreed upon, the paperwork filled in and posted off with a cheque. The large parcel duly arrived and we all eagerly grabbed our outfits to try on and gloat over. Sadly, they did not have my size and they sent another one in its place, same colour and pattern but in Mum's size! She got two dresses and I had none. We had to go shopping for my dress before going to the show. Coles sold clothing in their stores and this was where mine was bought.

Mum and Emmie had knitted us all new jumpers which was a

real treat and we each had a new pair of shoes. This year I got two pairs of shoes, both red but in a different styles. Boy, did I feel spoilt! Winter was a lovely time to have shoes. All up, we probably only went to three or four shows while on the road. Included in this was a circus in the very early years in Walgett.

As I took an interest in knitting, Mum taught me how to knit but for some reason I could not follow a pattern. I was left-handed and Mum taught me to knit right-handed, so I knitted to the casting off section at the under arms and Mum finished off the garment for me. I had a plain knitted jumper that year. I had been working on it for a while but I was a slow knitter, only doing one a year. It looked a bit uneven as I was not consistent in my tensioning of the stitches, but I wore it with pride and it was the colour I wanted – red.

The only ones to ever be punished in our family were Mary, Mike and me as the rest of the family were quite perfect! The hierarchy in the family was Les, Beryl, and Emmie who could never do any wrong, Mary was Dad's fourth favourite and I was Mum's, being favoured over Mary. So I did get spoilt to a certain extent, rather more so than Mike and Col who were the forgotten pair in the family. Col was with the stock all the time so no one saw if he mucked up. Sadly, Mary looked like Dad's mother, whom Mum hated with a vengeance and she was often unjustly hit for some slight that was only in Mum's imagination I am sure.

If Mum got annoyed at us for not doing our work properly or not quickly enough she would burn the book or magazine we were reading at the time. I found this very hard to tolerate, as our life was hard enough without being punished by destroying our only enjoyment.

When we were punished we had to choose our own stick. The hits were never gentle although as we got older, it was our pride that was hurt more than the whack. One day, out of the blue, Mum told me go and get a stick so I could be punished.

'Why?' I asked.

'You know why!' she replied.

'I don't know why, tell me?'

She refused. I was hit and felt it was unjustified. My feelings were hurt along with my bum. I turned to her and said sternly, 'I have not done anything wrong.' I then saw a look of doubt on her face. As I turned away, I could see Les smirking. I was sure that whatever the story was, it had come from him. Interestingly, later in the day Mum gave me two brand new tea towels to go into my "glory box". Actions speak louder than words.

If Mum was really in a bad mood, when Dad came into the camp she would tell him what we had done or said and he would also give us a hit. Mostly he just gave us a telling off or if he was in a good mood, which was quite often, he would treat it as a joke. If he felt obligated to hit us, it would just be a tap and pretend angry words to lighten the mood. Sometimes as "punishment" he would put the guilty party on his horse and send them out to the sheep with a made up message for the other riders, which was no punishment for us camp kids at all. Dad could normally rally Mum around into a good mood. She was a real enigma, a kind of a Jekyll and Hyde.

Cousin Ron Kemp saw and heard Mum telling one of us to get the stick to be hit with. He tried to pull her into line by saying we should not be getting our own bashing weapon, so she told another one of us kids to get it! Doing that was like cheering on at a hanging. None of us except for Les, got any satisfaction from seeing their siblings get hit.

We were getting close to Goondiwindi and it was one of the coldest winters I can remember. The black soil plains may have had something to do with it. The cold bitter winds cut through me and even on a sunny day, it was miserable and cold and I would go to bed just to keep warm. In the bitter cold our clothes did not stop the chill. As Mum drove the truck off the road to the camping spot, we found a huge load of overcoats dumped along the fence line. They were very good quality and of various colours. What a treat for us. We had a good rummage and got a coat or two each, which we really appreciated. We never did find out how or why they were dumped, we were just very grateful for them. We must have looked a sight, walking around in our long overcoats, but they kept us warm on those icy cold days and we used a couple of the very large men's overcoats as extra blankets.

We still had our penpals and as we were half way between Moree and Goondiwindi, Emmie and Shannon Holmes decided to meet up after so many years of writing to each other. Dad and Mum agreed and Shannon and her sister Cleone came and picked Emmie up. Two weeks later Emmie returned with Shannon's brother Bryon. She had gone into a stock and station agency and got a job on a sheep station looking after three children for the Matchetts and she was leaving home. Our parents agreed, so Emmie packed her meagre clothes, leaving most of them for Mary to wear, and the couple drove off into the evening twilight. This meant Dad was now a stockman down so Mary stepped into the role that Emmie had vacated.

We were now quite close to the Queensland border and Dad went to Goondiwindi for groceries, mail, fuel and whatever else was needed and came home with a funny but horrifying story. There had been a dance at the local hall and people had come from far and wide. In those days the children came too and as the kids got tired and fell asleep, the parents wrapped them in a blanket and carried them out to their cars where they slept contentedly as the parents had a merry time inside the hall. Someone, probably a group of young men, thinking what a lark it would be, went to the cars and swapped the children! Eventually, all the parents went home and as they carried their children in to settle them into bed, they discovered they had the wrong child or two! Phones began running hot as children were found and exchanged. Can you imagine this happening today?

We were fast approaching the end of the trip when we were caught in a mini cyclone that came through the camp and tipped the caravan over. We had gone ahead to set up the camp when we saw the storm building up. Mum pulled the truck up in the centre of a clearing and we could hear the wind as the storm came towards us.

Just before the wind hit us, Mum got a tarpaulin out of the truck and Mike, Beryl and I sat in the middle and she wrapped it around us. Mum stayed in the truck and turned the vehicle into the wind as the truck shook with its raging power. Suddenly, the wind changed direction and hit the side of the truck and shook it badly, turning the caravan over on its side. Fortunately it fell on the side the fridge was on and the gas tank was on the tow bar. If it had have fallen on the opposite side, the wall would quite possibly have fallen in with the weight of the fridge.

When Dad came along, he put me on his horse and sent me to get Col to come into the camp to help. Dad climbed into the caravan and pulled the fridge up to the door and Col hoiked it all the way out. They placed the fridge away from the work area and then emptied the caravan of other weighty gear. When Dad was satisfied he had it light enough, he put two ropes around the van, tied it to the back of the truck and pulled it back up onto its wheels. Apart from a bit of water getting in, no real damage was done. I think the cyclone was called *Little Audrey*. There were a lot of branches snapped off, trees knocked over and a couple of sheep were killed but no other damage in our little corner of the world.

The Moonie area has a great abundance of wilga trees and sheep love to eat the leaves. It was funny to see the sheep stand on their back feet to reach up for the yummy delicious leaves. If they could, they would place a foot on a branch to pull it down and their mates would all gather around and have a nibble with them. You could see where the sheep had been as the lower branches had all the leaves eaten off. These trees were an ideal way to drought-proof your property as you could cut off the branches to feed the stock. Wattle was also in great abundance in this area, growing in amongst the other trees and bushes.

We delivered the stock to Sid Derrick's station *41* in Moonie and from there we drafted them out into mobs – the Penfold brothers, Noel Murphy and Fraser Horn. We ended up taking little mobs of sheep on trips as the other properties were quite near. This would have reminded Dad of his very early day's droving in and around Wellington, New South Wales.

When we delivered the sheep to Noel and Pat Murphy's property *Donna Downs*, little did I know I would meet this delightful couple again some forty years later when I worked in a nursing home in Toowoomba.

As Fraser Horn did not own property, we drove his sheep around for a couple of weeks until he sold them in the local Meandarra sheep sales.

Now all the sheep had been delivered we could have a bushman's holiday. We camped for a while outside the Kelly's property, *Kinkora*, where there was a big reserve for the horses and the Moonie River running through provided ample water for both the stock and us.

At this stage, all we younger ones still could not swim. Col and Les were having a whale of a time swimming in the river while Mary, Mike, Beryl and I paddled on the edge. Col swam over and grabbed Mary and threw her in and after she gathered her wits, she started to dog paddle over to the river bank. Before I realised it, he did the same to me! I floundered, splashed around and started to sink, so he grabbed me by the hair and pulled me out. I was very frightened and spluttering and coughing up water. After I caught my breath, I called him every filthy word I could think of, all the while crying.

In the water there were dead trees, branches and roots, which we could easily have been caught on as we went under. We didn't ever tell our parents what had happened as Col would have been in serious trouble and for once, Les kept his big mouth shut! Mike being the little trooper he was, dog paddled out. To this day I cannot swim, even after several swimming lessons. Dad would say I could dive like a duck and swim like a stone!

Also camped on this reserve was an elderly couple who lived there permanently – Jack and his "Missus". They had a very old caravan under the cover of an old tin shed that stood under a few sad looking gum trees and they drove an old ute. They did a lot of kangaroo shooting in the area for the farmers and I quite often heard them shooting at night.

When we first arrived the Missus, who was very English, insisted on "mashing" us all a nice cup of English tea. She boiled the billy over the open fire, put the tea in the pot, poured the water in then twirled the spoon three times one way and then three times the opposite way and took great pleasure in pouring it into china cups with saucers. She offered milk and sugar, which we all accepted and we were expecting great things. How disappointing it was, as it was a brand we had not tasted before and weaker than we were used to. Dad said later, 'It was just a f…ing cup of tea, mate. It only had six tea leaves in it – the pot that is, not the effin cups at that.' In our camp only Billy Tea was ever used. Dad was a real connoisseur of tea. The billy had to go on the downwind side of the camp fire to boil. Dad could always tell if the billy had only simmered or had not quite got onto the full boil. If the tea was a bit weak he would sarcastically ask, 'Was there a wind blowing when you threw the tea leaves in?' We were never stingy with the tea and I cannot remember ever running out of tea. To settle the leaves to the bottom of the billy a tap on the side of the billy with a stick or spoon soon saw the tea leaves floating to the bottom. Dad had the art of swinging the billy over his head to settle the leaves and we kids learnt to do that too. Although we practiced with cold water in the beginning!

For a long while after our tea with the Missus, we would ask Mum in a posh voice to, 'Mash us some English tea,' or 'How about a cup of English tea?'

One day Mike found a bullet and gave it to Mr Jack to use. Mr Jack decided he had better try it out to see if it would work, so he fired it off and it worked! Their camp smelled very badly of stinking dogs, dead 'roo carcasses and cigarette smoke – their butts were everywhere. All the dogs were full of fleas and had mange. The Missus chain-smoked and she had a habit of tucking her hair back behind her ears so her hair on each side was a mustard yellow colour from years of nicotine being run through her hair with her fingers.

Mr Jack shot himself in the bum one day. The silly old codger used to throw his gun on the car seat and then jump in. This day the gun went off and it shot right across both cheeks of his bum with no serious harm done. The Missus made him lie on his belly and she used goanna salve to put across the gash until it healed. He was as tough as old boots. We were amazed that he never got an infection, as his car was really dirty and smelly and his clothes the same.

The Kelly's had a grown-up son called Michael who lived at home and loved a drink and party. Driving home late one night, he hit one of our horses and maimed it so badly it had to be put down. The car was also damaged and Michael broke a leg, but luckily the horse was not one of the "good" ones. Both Mr Kelly and Michael felt really bad about the accident and Dad muttered about useless drunk drivers for months afterwards.

After a nice long spell there, it was getting near Christmas time so we packed up and took all our goods and chattels back to Moree. Dad's droving friends the Duvalls lived on the outskirts of town and we left the dogs with them and our extra gear in their shed. We travelled on to Dubbo to spend quality time with Dad's mother Emily Thomas and his sister Anne. Cousins Fred Pout and his sister Juanita still lived at home. Our caravan was parked in their small backyard and we spent about two weeks there. In the yard there was an old broken down car where cousin Fred did most of his courting! He was seeing a young woman called Ruth who he later married and he sometimes took the older kids out with him to show them around a bit, but us three "young ones" had to stay home.

Five adults and seven kids lived in the one small house. We all still slept in the caravan and the back of the truck, except for Mary who shared a bed with Juanita. The real treats were having a toilet that flushed, a bath with fresh water and towels that were soft and fluffy.

We met up again with Dad's brother Fred, his wife Lil and their daughter Shirley. Uncle Fred had a used car yard in his backyard where he did all his wheeling and dealing. He did a brisk business with the local Aborigines and the poorer class people. Some years before this, Dad had bought his first motor vehicle from him. Dad had asked Uncle Fred, 'What's the guarantee on it?' and Uncle Fred had replied, 'Guarantee finishes once it's out the yard!'

As a treat while in Dubbo we went to the swimming pool a couple of times and the Saturday afternoon movie matinee. The trailer of *Tarzan* was always welcomed as who knew what he would do next? The rest of the time we stayed at home and lazed around – very boring as we were bush kids and wanted to be out doing things. Mum told us of her first experience of going to the movies and seeing a cowboy star she loved called Audie Murphy. There was a train in the movie and it seemed to be coming out of the screen as it got larger and faster. Mum screamed and ducked under the seat – that story tickled us no end.

We all learnt to ride a bike that summer and that was wonderful but we were not allowed to ride too far. Juanita had a part-time job at a service station and was friends with the owner's daughter, so if we happened to be around we could have a soft drink for free, which for us bush kids was quite a treat.

Fred had a daily bread round and the horse knew the route so well, Fred would walk and deliver the bread to the houses as the horse plodded along and stopped and started on command. I thought that was marvellous as our horses were not that clever. Fred always made sure their house was the last on the round so we got all the leftover bread. We had French sticks, Cobb loaves and sweet buns, all breads we had not had previously. We all enjoyed this "hot" bread. Though the story was that Fred was allowed to give it to us as long as he could not sell it on his route.

My aunt and uncle had a television set but no colour in those days. We were only allowed to watch it when the grown-ups were watching it, so we mostly got news and boring stuff. We still thought it was great, as we had never seen one before. I am sure there were children's programs but I cannot remember them, only having to be quiet so all could listen. Aunt Anne had sensitive hearing so the television was always turned down low – so low we had to sit close to be able to hear it. Fred would often turn it up for us very slowly so Aunt Anne did not notice. If she realised it was getting louder, she would ask Fred, who was the person who changed the channel or dealt with the sound, to turn it down. He'd pretend to turn it down and she would reluctantly agree that it was low enough. Fred would give us a sly wink and the show went on. Aunt Anne was a softie and would often insist on putting the telly on a "special" show for us kids to watch and we were grateful for it. Thinking about it now, that was when we would be at our quietest.

BOURKE 1965

Traditional land of the Nyemba people

Early 1965 we did our last trip from Bourke to Meandarra Queensland. The stock buyers this time were Sid Derrick, Val Coggan, Bert Wormal and Dad invested in his first mob of sheep. Between them they bought another flock of just over 8,000 sheep. Bruce Knobbs, Ron Hunter, Sid Derrick and Dad travelled to the sales to select the sheep, Dad taking his trusty ready reckoner that he rarely went anywhere without. This time the offerings of sheep were a bit thin on the ground and we had sheep from various stations. This droving trip went pretty much the same as the other trips, starting in summer and ending in spring – February to October.

Dad traded a horse for a new dog, Snow, that showed great potential. Snow would madly run everywhere and tire himself out. Dad placed the dog's front paw into his collar to slow him down a bit. It worked well. When in burr country, Dad made small leather booties for the dogs so they did not get sore feet from the burrs.

Early one morning after the men and sheep had left the camp, we packed up and Mum was driving the truck, pulling the large caravan up the steep bank on the side of road. The curvature angle of the road was very high and the truck and caravan tipped to a dangerous angle. Mum panicked and thought they would tip over. She stopped the truck, yanked the handbrake on and yelled for us to get out of the back of the truck very carefully, so as not to tip the balance. We could see dust in the distance, the first sign of a vehicle coming and eventually the car came into sight. Mum flagged it down and as it came to a full stop, the dust billowed around the car and Mum. It must have been a Saturday as it was full of footballers and they gladly passed a message on to Dad. He cantered back to his stricken wife and kids and managed to get the truck and caravan onto the road without any mishap.

As Dad became more successful in his profession, he found sleeping at night more difficult. He and Mum would smoke several cigarettes during the night. When Mum got out of bed in the mornings she would have a coughing fit and light up a cigarette as she roused us out of bed to start the day. She would quite often look at the cigarette she was smoking and make the comment, 'These bastards are going to kill me one day.'

We had pulled right off the road for a dinner break and camped in a bunch of trees. This gave great shade for the caravan and to have dinner under a good dense gum tree and its cool shade made a nice change. When I needed to go to the toilet, I tore a page out of the magazine, *Australasian Post,* for toilet paper and meandered off to find a good sized tree trunk to squat behind; not that I needed a big tree to hide my skinny bum. The trouble was, if a sibling caught even a glimpse of a skirt hem, you got a cat call, wolf whistle, a stick thrown at the tree and for total humiliation, you would hear the shout: 'I can seeee you.' This also let anyone else in the universe know what you were doing. Privacy, what privacy? I would walk to the opposite side of the camp where the stock was coming from. I'd be safe from view from the riders with the stock, even though I could hear the sheep baa baa-ing and possibly dogs barking.

I spied a likely tree, had a quick glance around and appeared to be in the clear, so I pulled my skirt up, pants down and squatted. What a relief. Next moment, I heard a noise. I glanced around and nearby saw a dog followed by a man nonchalantly riding along, looking straight ahead. Did he see me? More than likely, but not by a flicker of an eyelid did he let on. I prayed to be struck down dead, there and then. I stood up after he passed and straightened my clothing and he kept riding into the camp. Mum and the rest of them came out of the caravan for a chat with him while I stood behind the tree, waiting and waiting for him to ride off into the sunset. Bloody boundary riders! He stayed chatting for ages and eventually Mum realised I was missing and started to call to me. 'Pat, Pat, Paaatsy, where arrre you?'

Oh no, I had to go and face him. I prayed for the ground to open up and swallow me to stop this total humiliation. The rider had his back to me and I stood in front of the tree where the family could see me but he couldn't. Eventually, he rode off towards the oncoming stock. I don't know if I ever met this person again. My hope was that he had only seen me bum and wouldn't recognise me face!

This lunch break we had no cold meat for dinner so Mum boiled up packets of chicken noodle soup and we girls sliced and buttered some bread. Dad was first in to the meal break and sat down, looked into the bowl of soup and scooped up a few mouthfuls. He stirred the soup thoughtfully, looked at Mum and asked, 'What's this?'

'Chicken soup,' Mum said.

Dad replied, 'Did the bastard walk through it on stilts?'

Mum immediately lost it and said, 'Go without then, you ungrateful bastard!'

We thought this was a huge joke and along with Dad, all roared with laughter, although Mum was not amused. We were lucky she did not throw it over us.

Life at home was getting intolerable for Col with camp fights and arguments becoming more vocal and uncomfortable. It all came to blows one night during dinner. We were sitting around a dead tree that was lying on the ground. Dad had lit it in the middle so there was a good glow showing the whole campsite and surrounds. Dad had been abusive to Mum over the last week or so. Col told Dad to leave her alone as she was sick and the rest of us were helping as much as we could. Dad jumped up and went to punch Col, but Col was on his feet just as quick and ready to fight. They faced up to each other, took their mark and traded punches around the camp fire and out into the dark. Between Col's youth and skill and Dad's old age and treachery, no one won the fight, they were a perfect match. A few days later, Col told Dad where he could stick his job, asked for his cheque and within an hour he had hitched a ride into Goondiwindi. We were now down to three horsemen, Dad, Mary and Les. I was still needed at the camp as Mum was still taking fits on a semi-regular basis and I now did the majority of camp work, including the cooking and setting up the sheep break each night. Mike and I now shared being the extra person with the stock, which we both loved doing. Mike also proved to be my trusty camp helper and we managed the best we could.

I now stood 4 feet 11½ inches tall (151 cm) and I could carry the steel pegs four at a time. Mike was now thirteen and helped me, but I still had to stand on a kerosene drum to hammer the steel pegs into the ground. I would weave the pegs into the fence wire. With me pulling and Mike pushing the peg, we made it as tight as we could and then with Mike pushing with all his might, using both hands I lifted the hammer up and banged it down the best I could onto the top of the peg. This had to be done quite a few times to get the peg into the ground to make the break firm enough to hold the sheep in. Mike would close his eyes and I think we were both grateful when he could step back from the peg and my wild swinging of the heavy ungainly hammer.

Standing on an empty kerosene tin that was not very stable and wobbled around a bit made it difficult to hit the right spot on top of the iron peg. If the ground was not too hard it was okay, but it was murder in the black dry soil. We managed okay and I did not have to do this every day, but quite often Mum was not capable of doing it. If Mum felt a bit better later in the evening before the sheep came in, she would pull the pegs out and tighten up the wire and re-hammer them in. If she could not do the adjustments, Dad would do it after the sheep were safely enclosed in the break. As time went by and I became stronger and more adept at doing the job, the fence line improved in firmness and no adjustment was necessary.

The job was reversed the next morning when we had to pull the pegs back out again and roll the wire. This became my job and I had many a piece of bark off my arms through rolling the wire. Once it was rolled a few times, I had to stand and give it a tug so the roll tightened up otherwise it was so big, it would not fit in the back door of the truck.

Quite often before riding off, Dad would walk around the break and tug the pegs out for us, which we were ever grateful for. Mum would potter around the camp, packing up as much as she could though it was still her job to drive the truck. She never had an inkling when she may have a fit and none of us kids picked up signs either. Dad would often get the stock on the way and then canter back to the camp and help do the final packing and Mike or I would then take his horse to the stock. By now it was smoko time so the truck was pulled up on the side of the road, a small fire would be lit, the billy placed on the tripod and if Dad was hungry, chops would be grilled on the shovel, which we all enjoyed.

Emmie wrote to Mum telling us that she and Bryon Holmes were getting married in Goondiwindi. Luckily, the wedding was after the droving trip was finished, so the excitement of the first family wedding began.

Sid Derrick drove out for a visit to see how the stock were travelling and he told Dad to keep the stock moving until his oats were ready for them, so when they arrived he could fatten them up and sell them. It was the feed gathering trips that we liked so we could have a few easy days here and there and if there was plenty of feed, the stock inspectors did not expect you to travel long distances each day. A day or three on a reserve was a "bushman's holiday" for us. If lucky enough there would be a grid at each end so all the stock could be let go. Our big job would then be to water the dogs but surprisingly, after a few days, instead of relaxing and enjoying the change of pace, we would be ready to move on. These Bourke trips lasted 6 to 10 months just poking the stock along. We were never in a hurry as Sid didn't want us to arrive until October, when his oats were high and lush.

We met a lot of kangaroo shooters on our droving trips. Some were out to cull and the professional shooters killed for both the meat and leather from the hides. The 'roos could get to plague proportions and then something had to be done once the dry season began and they competed with the stock for food. The recreational roo shooters were the most dangerous. They shot anything that had eyes: horses, sheep, cattle even humans could get killed. When we were in the vicinity of the shooters, we always had the fire burning brightly so they knew that there were people around.

Most of the shooters would pop over for a yarn and ask where the stock was so they could avoid that area, which was always appreciated. In those days, the shooters drove into places and started to shoot. Most station owners preferred to choose who they had on the station so they knew who they were and that they could trust them to do the right thing. Another problem with the 'roo shooters was that they would sometimes leave gates open and the stock would get mixed. If they did that too often they were banned. Dad did not like them much and he used to say they should "get a real job". The shooters stank of stale blood and other smells. After shooting a 'roo, its throat was cut and it was hung upside down on the side of the truck, so it would bleed out. It was then taken to the nearest chillers to be sold.

Dad told us of his misdemeanours as a youth in Wellington, New South Wales. There were dances at the local hall and as the adults danced, drank and ate, the boys would be mucking up outside. One of the things they used to do was unhook the horse from the sulkies, put the shafts through the fence and take the horse around outside the fence and reharness the horse into the shafts. The owner would come out, get into his sulky and try and drive off and get nowhere fast. Dad and his friends and would be standing off in the dark having a good old laugh as boys do. If anyone made the mistake of leaving any bottles of alcohol in their carts, Dad and his mates would all have a sip of whatever was in the bottles. It did not matter what it was, they drank it.

Dad was starting to go a bit soft… On the grocery runs into town, he would bring back a tube of condensed milk or a tin of pineapple juice and give one to each of us while we were with the stock, rare treats indeed. We had some good times on these last trips. The extra money made our life a little easier and the quality of food improved – Mum bought more packet food, which we all enjoyed. We camp kids had a couple of long pants each and long-sleeved shirts and good shoes – with socks! Our parents fought a lot more than they did before or perhaps I was getting more sensitive to the arguments that were happening on a near daily basis. Mum was more suspicious of Dad's trips to town and the people he was bringing out to the camp for meals. If there was a couple, more often than not the women were attractive and flirty and Dad gave them heaps of attention. Mum's disgust and jealousy showed but it did not appear to matter to Dad – he was just having a good time.

The stock was dropped off at Sid Derrick's *41*, now called *Moorooka*. Once again, we drafted the sheep there, Sid's happily went into his fully grown oats. We delivered Bert Wormel's to his station, *Warringa Downs*; Val Coggan's were driven to *Enarra*, his station; and Dad's mob went into the Meandarra sales. Sadly, Dad's runts were passed in and we had to drive them around trying to find a buyer. This was the last trip to Bourke as the price of wool had fallen drastically and no one was interested in any great mobs of stock anymore. By this time, trucking stock around had become popular and some drovers bought a truck and started their own business.

Not long into droving our own mob of sheep, Dad received word from the local police that his mother, Nanna Thomas had had another heart attack and was not expected to live. He arranged agistment for the stock on a property called *Wilga Park,* owned by Bert Meackle outside Westmar.

The parents took Beryl with them to Dubbo and on the way past Boomi, they dropped Mike off with Emmie's fiancée Bryon. That left Les, Mary and me at "our" reserve, near the Kelly's *Kinkora*. We camped here for about two weeks or more while the parents were away.

Mr Kelly called into the camp to make sure we were doing okay and to ask if we needed anything. We offered him a cup of tea, which he accepted and we offered him some Sao biscuits, which he declined, saying he never ate them. A week or so after this, we were at his house and his wife gave us a cup of tea and Sao biscuits with tomato and cheese. Mr Kelly ate them with gusto and I said to him boldly, 'You said you never eat Sao biscuits.'

'I never eat them while I am out as they crumble everywhere and make a bit of a mess,' he said kindly.

I thought he must have been worried that the ground around the campfire could not handle a few crumbs! It was not until my later years when I learnt some etiquette and the finer points in the art of eating in front of others that I knew what he meant – after the first bite they crumble into several pieces.

While Mary and I were at the camp on our own, a grader driver drove up with his full complement of work gear, this being the grader, car and caravan and he camped not far away from us. We decided to invite him for dinner as he would have been lonely for company, just as we were. When the parents got back, did we get into trouble for it! Mum and Dad said we shouldn't have spoken to him. Although it was okay for Dad to cadge fuel from him! We enjoyed this pleasant man's company as he was easy to talk to and treated us like adults. Grader drivers lead a lonely life working alone on the ungraded roads in the outback, never seeing anyone from one week to another. About once a week they move camp. Most of the time under a tree, just off the road. This man told us he would grade himself a nice spot to camp on, removing the grass so he could light a fire if he wished.

Our grey mare Penelope, got a large stake in her foot while the parents were away. Richard Meackle, the handsome son of Bert, lived across the Moonie River where our sheep were, and offered to get the stake out of the mare's foot. Mary was holding Penelope's head and as the stake was pulled out, blood gushed and Mary fell into a dead faint right at the mare's head still holding the reins. It was the first faint I had ever witnessed and what a shock we all had – including the mare, who jumped back. Richard bathed Penelope's foot with Condi's crystals and bandaged the leg up and all went well – both ladies made a full recovery. And who did this young man find rather attractive? It was I, the brunette and not the red head who I thought he'd prefer.

Richard drove his rusty, dusty old ute into the camp one day and told us some of the sheep were flyblown. The three of us had a round table discussion and the next day Richard, Les, Mary and I went riding looking at the sheep. The paddock they were in had been pulled and the trees were still lying on the ground with huge holes. If the sheep were flyblown, they had a tendency to hide in these holes or up against the logs or under the logs. If we saw a sheep lying down, we knew that it was flyblown, as when they were well they would try to run away when they saw us. We carried our trusty bottles of KFM on the side of the saddles along with a pair of shears so after we snipped the wool away we could soak the offending section of the sheep's hide. Having our trusty dogs with us who would pull a sheep down if they tried to get away was a great help. Quite a few sheep were lost as it was not possible to find them all in the large paddock. I hate to think how many died a long agonising death in the heat of the summer with the pain of the maggots crawling all over them and slowly eating their flesh away. A few we had to hit on the head as they were too far gone to survive. It said a lot for mulesing, which I had not yet come across at that stage. (Mulesing is a process where strips of wool bearing skin are removed from around the buttocks of a sheep to prevent them becoming flyblown.) After we had control of the flies, Richard still had his work on his father's farm so Mary and I rode out every other day and we kept on top of the problem. Ideally, the sheep would have been brought in for crutching. Les, ever the pain, had spied wild pigs in the paddock with the sheep and he decided he would rather catch them. He cut the handle off the broom and tied a butcher's knife to it and off he went, chasing wild pigs. He would gallop up beside the offending pig, stab it until it dropped, jump off his horse and cut the pig's throat. He was useless to us so Mary and I went about our business without him.

The parents were away for two weeks. Nanna survived the heart attack and they found a nice unit for her to live in near Aunt Anne and she settled there quite happily. When they arrived back at Westmar the sheep had eaten out the paddock at *Wilga Park*. Dad managed to get a lease on the reserve where we were camped for several weeks so each day we rode around the sheep to keep them quiet, so they would not rush off upon seeing a horse or dog.

Michael Kelly was doing a bit of sheep work and asked us if we would like to go and help him out as he was still limping badly from the leg he had broken the year before. We agreed and rode our horses to his stockyards the next morning, dressed for a day's work. He was planning on running his sheep through a shower dip and was not quite ready for his helpers, so we meandered around the yards while Mr Kelly and Michael tried to get the pump to work. We were walking through the shower race when lo and behold the pump started and we screamed as the cold, dirty, smelly water drenched us, much to our consternation and Michael's delight. Luckily, it was a warm day and we dried off soon enough. So much for trying to impress the eligible bachelor! Mary and I laughed about this incident many times over the years as being one of the most embarrassing things to happen to us. Around this time Dad realised he had better start keeping a closer eye on his daughters and many a young jackeroo was told to go on his way and be quick smart at that.

While we were at this reserve, Emmie's wedding was upon us and we went to Goondiwindi for her big day. We were all in a motel and poor Emmie had to help us girls get dressed and do our hair as we did not have a clue. Emmie looked stunning in her white dress and long red hair flowing over her shoulders. Mary and I were bridesmaids and we wore light pink frocks, low-heeled shoes and Beryl was flower girl and wore a light blue dress. It was the first time any of us had been in a church. The ceremony seemed to take forever and both Mary and I found our new shoes very painful to stand in for such a long time. Being inside a church was interesting and the next interesting part was the wedding reception. What a fiasco! Out of fifty wedding invitations sent, we had only our immediate family, Ron Kemp and his wife Mary who had brought Nanna Thomas from Dubbo with them, and a handful of Bryon's family and friends. Dad and Mum had sent the invitations and they had not heard from any of the guests, so Dad assumed they were all coming. Come dinner time, meal after meal after meal came out and was plonked on empty tables. I was humiliated for them, such a disappointment for the happy couple. Emmie was devastated but put on a brave face and tried to enjoy her wedding day We could tell it affected her but no one else would have picked up on it. Mary and I loved the attention we receive as bridesmaids, all dressed up in our finery and looking beautiful – or so we thought!

DALBY 1966

Land of the Burrungum or Jarrawah Aboriginal people

After the wedding we took the sheep back on the road again, travelling from Moonie to Dalby via Tara, Chinchilla, Jandowae, Bell and Dalby. The Dalby area was not set up for travelling stock. The lanes were small and the grain farms were either poorly fenced or had no fence, which caused havoc with trying to keep the sheep on the stock route and not in the farmers' paddocks where they much preferred to be. The sheep had not seen such lush green grass in their lives and with the greatest of ease they managed to get into nearly every paddock we passed. It was okay for the men as they managed to keep the sheep under control with their dogs, but when Dad went to town and left the camp kids in his place, some of the dogs would not work for us. Here we would be madly running after the sheep with a bush in our hands, trying to get them back from where they entered but of course they could never find that particular place again. They were having such a good feed and bellowing so loudly the others would come along and try to join them. If a dog owner was not too far ahead, he could send his dog back but Les, being the contrary person he was, rarely did this, much preferring to sit back on his horse and enjoy the show. If we had a go at him in front of the parents, he would deny knowing what was happening.

In 1966 pounds, shillings and pence turned into dollars and cents to fit in with the American dollar system to make it easier for common trading. So $2 equalled £1, $1 equalled 10 shillings and so on. It was confusing for all and there was a suggestion to not bring in the system until all the old people died...? I am not sure who the bright spark was who suggested that one! For the first year we could use both currencies and then we had to say good bye to the old pound and hello dollar.

On the way to Dalby on the Moonie highway before we turned off to go to *The Gums*, a young man in a utility pulled up asking Dad if he had any horses for sale. Dad gathered the horses in a bunch and he gave this man, John Davies from Tullumun, his choice of any horse to buy. He chose my beloved Gayboy. He was sold for $100. Dad did not tell me until the next day. He emphasised that it was a good price and as it was pointless for me to say or do anything, I did not comment. This had become my normal way of communication. If you kept your mouth shut, you would not get into trouble. I soon found a quiet spot under a large gum tree and sobbed my heart out. I loved that horse dearly. How could Dad have done that to me?

As we were entering into the Dalby area, we were travelling along the road near a village called Macalister and who should stop on the roadside and have a chat but a man we met in Westmar, Bert Harris. We were quite close to his parent's farm, *Argyle*, between Macalister and Dalby and whenever he saw us he stopped to chat.

We were at a hall between Bell and Dalby and the fence was a bit off the road and not up to standard. Nearby was the fenced hall so Dad decided to put the sheep break on that, he was halfway through putting it up when a car drove up and the driver told him that he could not put the sheep break there as the following night there was going to be a dance. Cars would be using the space to park and the people would be walking through into the hall. Dad belligerently told him that it would not hurt them to walk in sheep shit! He eventually gave in and moved the lot over to the fence nearby.

We were getting nearer to Dalby and Dad and Mum had gone to town. Mike and I were at one side of the mob sitting on horses having a chat and a young man drove up and had a talk with us. After a long, enjoyable chat Mike invited him to dinner. I looked at Mike with a worried frown as I knew Dad did not like us inviting just anyone to the camp, especially a strange young man. Mike didn't tell anyone he had done this and being gutless, I did not say anything either as I knew it would mean trouble either way. Dad had to kill a sheep that evening and half way through the job, the sheep was upside down, hanging off the gamble and the young bloke drove up, all dressed up in his best clothes. I had made myself scarce as soon as I saw him coming as there was lots to do. Mike ignored his guest and Dad turned to the young man and asked, 'What do you want?'

He answered, 'The young bloke invited me to dinner.'

Dad looked at him and said, 'You are now uninvited so piss off.'

When confronted, Mike denied he had invited him and I said nothing. I really felt bad for the poor fella and regretted not contradicting Mike's invitation at the time. I knew Dad would not be pleased about it.

Dad met up with a young farmer, Frank Bach and his sister Jeannie, just out of Dalby on the Blaxland Road. The Bach's harvesting had been completed and Dad and Mum had to go to Dubbo on some business. Dad asked Frank if he could agist the sheep into his paddocks until he came back as we kids were not licensed to drive to keep the stock moving on the stock route. Frank agreed to this. We camped in the stock route at their front gate while the parents were away for over a week. Dad and Mum took Beryl with them so that left the four of us to look after the dogs and horses in the paddocks with the sheep.

Mary and Frank got on well and Les got on really well with the manager Barry, who lived in the house with the Bachs. Each night we played cards, watched TV and had a holiday ourselves. Before the sheep could be left safely in the paddocks we had to do a bit of fencing to keep them in and fix any holes they managed to find. Even though there was green feed for them on the farm, lush green grass grew in abundance beside the fences. As they were chewing the grass, they just followed the feed line and walked through the fence into the next paddock.

While the parents were away, the Bach's harvested another paddock and with great gusto we "helped". I enjoyed working in the kitchen with the cooking and baking and taking the meals out into the paddock to the men. We had a picnic morning, afternoon smoko and dinner, which made a change from the humdrum life on the road. Not moving for a whole week was bliss and, when the parents came back, there was still plenty of feed for the stock.

Dad, Frank and Barry went to town late one evening and they all got on the grog in one of the pubs with a couple of the local cops. Dad must have been having a great time as he got drunk. Doing the right thing but not sure by who, the two cops escorted Dad and Frank to the edge of town and told them to get home safely as they had to start their shift.

The nights in the pub became a regular event and for some reason Mum sent Mary and or me with them. After a particular evening on the grog, Dad decided it was time to go home. Not far out of town, Dad's drinking caught up with him and he could barely stay on the road. Mary and I had to tell him when he was in the middle of the road as that was about the only time he was on it. We girls thought it was a great laugh, which made Dad play up a bit in the truck, but eventually we got back to the Bachs safely. This occurred a few times while we were camped there. I was never sure why either one of us got the job of going with him as we could not control him in any way at all. If the shops were closed we just sat in the truck waiting for Dad to come back out to us. Occasionally, Dad or Frank would bring us out a glass of lemonade or orange juice, which was gratefully accepted.

There were heaps of pig farmers in and around Dalby and Dad became quite interested in these piggeries. He was considering having one for his next adventure as droving was quickly going out of style.

Frank and Dad drove into Dalby one day and Dad drove back in a brand new green Toyota Crown station wagon, which was a great thrill. We all loved going on trips with Dad when he had to drive along looking at the feed and water for the stock. When we reluctantly left the Bach's farm with the stock, Mum still drove the truck and Dad drove the car, which proved not successful. He was needed with the stock so he had to get a new driver – that turned out to be me! The driving lessons began and what a shouting, crying match it became – he shouted and I cried. He made me so nervous I could not remember the simplest instructions: foot down, put in gear, foot on accelerator and drive off – car hop, car stall. Okay: start car, foot down, put in first gear, hop, stop. I would be driving along quite happily and the next thing my mind would go blank. He decided I had to learn to drive up a hill, but the only hill around was a steep creek bed. I drove down in the bottom of it, we were well hidden from oncoming traffic and the car stalled. He started to panic in case a car came along and of course, I panicked in case I got a hit over the head for my stupidity. With the stress of a car possibly coming along and smashing into us, I was so nervous I kept stalling but then I got it and drove up and out of the gully. Thankfully, the end of car lessons!

The next morning, after helping to pack the truck and caravan with the gear, I sat in the car and my mind was blank. I was thinking: *How do I start this thing?* Mum walked to me sitting in the car and asked me how I was.

I looked up at her and said, 'What do I do?'

From the outside of the car she told me the gears to use and away we went. I learnt to drive in one day! My driving consisted of crawling behind the truck and caravan and I was happy with that! I got more adventurous about three years later when I got my driver's licence.

Eventually we had to leave the Dalby area as the stock inspector started to get complaints about us. The farmers' crops were coming to a head and the sheep were proving impossible to stop from getting through the fences. The lanes were very small and the sheep would spread right along the road and the farmers had to slow down to drive through them. We were taking the sheep from Dalby back to Westmar looking for agistment for them. It was very dusty and Dad's old dog Speck was getting a bit blind and deaf but he was still determined to do his bit. As the sheep were going along, one of them broke from the mob and took off running with Speck not far behind. The sheep ran into a fence and Speck ran after it following the sheep by sound and smell. He ran smack bang into the fence. He was a very loyal old dog and he was worth more than two workmen.

One day as we were droving the stock along, Bert Harris once again saw us and stopped for a chat. Dad said he had to shear his sheep and Bert offered him his shearing shed at *Brenda Park* out of Westmar. *Brenda Park* did not have a house on it, just the shearing shed and some yards, so we camped nearby on the property and Bert arranged for his team to come in and shear the sheep. After they were shorn, Bert offered to buy them so Dad counted them into his paddock and we brushed our hands and walked away. We were pleased to be rid of them, as Dad was finding it expensive running the camp with no wage coming in. Dad sold off the horses and the dogs, keeping six horses for our use and a few of the good dogs. We moved back to Moree in New South Wales. Little did we know Bert would become a member of the family three years later.

CONDOBOLIN 1966

Traditional Land of the Wiradjuri or Wirrajuwrruy people (different spelling)

In 1966 southern New South Wales was in drought and no droving was available and we camped once again on *Ticklebelly flats* (*Lovers Lane*) out of Moree. Dad decided it was time to start out on his own and he drove the truck to surrounding towns buying cheap cattle in sales. Over a period of a few weeks he bought 300 steers, heifers and cows in calf. For a few months we drove them around Moree, Narrabri and anywhere they could get a feed. Dad would chop down big branches off large trees and drop smaller trees to give them something to eat. They were very poor and getting poorer. After Dad chopped down trees and branches over a few days, the cattle caught on that the "chop, chop" meant food. We had to try and keep them away from the falling trees and limbs. With them being rather quiet, it was difficult as they were not at all scared of us. The dogs did a far better job at it, but then the cows just stamped their feet at them and went wherever they wanted to. Never tell me that cows are not smart!

Dad had enough of using the ungainly axe, finding it hard work, so he bought a chainsaw. One day as he was cutting down a medium-sized tree, Mary, Les and I were trying to keep the cows from getting too close to it and as the tree fell Mary called out 'Timbbeerrrr.' Dad went off his brain at her for calling attention to what he was doing – as if no one would have heard the sound of the saw as its motor roared. Mary and I had a good laugh over that later on. Men!

We befriended many people on that trip as we needed all the help we could get. At one particular place in Moree – I will call it *Success Park*, the managers taking care of the place became friendly with the family. Dad was a real talker and charmer when he wanted to be and he managed to get some hay off this couple for the horses. He was also friendly with another bloke, Arthur, a few miles up the road, who also had hay and he would beg hay off him too. One day Arthur asked Dad, 'Aint the Sullivans giving you any hay?'

Dad replied, 'Nah, the tight bastards never give us anything.' Then he would go to the others and say the same thing, so he was getting quite a few bales of hay from sympathetic station owners or managers. Even station hands were happy to drop off a bale or two as they drove past on their way to town in their ute. I guess, like us, it was a bit lonely for them on this beaten track of ours and any excuse was good enough to stop for a chat.

The manager of *Success Park* was a real character. His wife would drive the ute and he would stand on the back of it and kick the bales of hay off as she drove along slowly. She could never keep the vehicle going smoothly and he fell off a few times, hurting himself enough to claim compensation. He used to say that as he fell off the ute he would call out "compo", so we were not sure how genuinely hurt he was but he made us all laugh at his antics. Sometimes sympathetic land owners would let us put the cattle into their stubble until it was all eaten out. That made our life more bearable, and the stock also felt better for this treat, just grazing and not having to do the ten miles a day walking to the next stop.

Luckily, there were bore drains in abundance that were still running so we managed to water the cattle and horses. The lack of feed for the cattle was the real problem. As the cows calved, we tried to save as many as we could, and handfed the ones that were rejected by their mothers. It was an ungrateful, unrewarding job as most ended up dying on me. It was also my job to lift the calves into and out of the truck. I was still quite small and only weighed about 90 lbs (40 kilos) wringing wet so it was a difficult job. If I was lucky and the stock was late getting off camp, Dad or Les would lift the calves into the truck for me. It was easy getting them out of the truck as Mike would bring them to the back of the truck and lay them down and I would grab them, and with their weight helping me, ease them onto the ground. If we had too many calves, Dad would sell them to anyone who wanted one. I was always pleased to see them go as it meant less work for me and I did not like to see them die after all my hard work. If a calf did not get its first drink from its mother, it meant it did not get its first taste of colostrum which was needed to line its stomach, so we put a raw egg in their first drink of milk as it was supposed to help.

If the stockmen did not keep an eye on the cows, a cow would calve and hide it under a bush and she would keep grazing along until later in the evening when she would then trot back to her calf. When it first happened, the men did not realise what the cow was trying to do, so they had to let her go and just follow her as she trotted along bellowing for her baby. The cow would go straight to the calf, which would still be sleeping soundly and until the calf had a drink, the cow would not move. After the calf had a drink it would possibly walk with its mother and if not, the stockman would place the calf over the neck of the horse and hold it there while he rode back into camp with the cow following along closely. It was not only the newborns that were hidden. The older calves would do the same thing, stopping for a rest while the mother grazed along until she decided she had better go and pick it up. After a bit of experience, the men knew which cows to watch to make sure they still had their calves walking alongside them. When possible we would have longer smoko and lunch breaks so the calves could rest a bit and keep up with the mob.

I was walking behind the cattle and I got a rusted nail in my foot. I levered myself back into the saddle and rode into camp for Mum's ministrations. The goanna salve did not work its magic and the foot became badly infected, so Dad took me to a doctor in Moree. The doc gave me penicillin and I then got a red itchy rash all over my body. So off to the doctor again. He gave me a check over and pronounced I was either allergic to penicillin or I had scarlet fever. Like a leper, I had to keep my distance from the family but I still helped to carry the wood and do other chores around the place. Eventfully I became better and everything cleared up. No one else got the rash so I guess it was not scarlet fever as there was no way I could be fully isolated from the family.

Eventually Dad decided to take the cattle to Condobolin where there was more feed as that area had received some rain. There was no feed in the stock routes between Moree and Condobolin, so the stock had to be trucked. First the cows were put in a sale to try and sell some. Those that weren't sold were trucked out.

Mum was not doing so well on her medication and she needed to go back to the Sydney hospital to have more tests done. Mum and Beryl caught the train to Sydney a few days before we had to load up the stock. Once again they stayed with cousin Ron. The car was left covered in the backyard of an old droving friend.

We had four horses and the calves in the back of the truck. I deemed it too early to feed the calves before we left for Dubbo and by the time we arrived they were very hungry and had to be fed late at night in the back of the truck. It was rather awkward as they all wanted a drink at the same time. Mike held a light as Mary and I fed them with their bottles. We were as grateful as they were for the food as it stopped their bellowing. We spent the night at Dad's sister's house. It was great to catch up with Aunty Anne and Nanna Thomas again as we had not seen them for a year or so. Cousin Fred had married Ruth and was now living in Sydney and had a couple of children. Juanita married Billy Boland and also had a couple of children. This stop in Dubbo made a good break in the trip for us as we had all been working very hard with loading the stock on and off and the day-to-day work in the camp. When we arrived at Condobolin the next day, the cows were in the sale yards as arranged.

We were in Condobolin over two weeks before Ronnie brought Mum back with his wife and the youngest three of their seven girls. We were expecting them so we had a nice big fire burning on the side of the road where we were camped. Mum had a wig on for disguise and pretended to go off at Dad for having such a big fire going on public property. We enjoyed these city folk out in the bush with us. Ron's three girls, Helen, Rhonda and Sharon were younger than Beryl and we enjoyed putting them on the horses and teaching them our ways. We showed them how to sit on their heels, carry their plate of food without letting the food drop off the plate and make toast by holding the bread over the fire with a stick.

They burnt the bread a lot of the time but we all had an enjoyable time. It was a real treat to spoil them. When Ron and Mary went back to Sydney, they left Helen with us as a playmate for Beryl. She was a sweet child and two years younger than Beryl.

The cows had a real trip as we travelled all over the area with them: Lake Cargelligo, West Wyalong, Forbes, Ungarie and the surrounding districts. Chicken farms were quite prevalent in this area. If we were close by, one of us would either walk or ride into the farm and buy a few dozen eggs, by now we were all eating eggs for breakfast. We were rich indeed. Buying them from farms was not cheaper than buying them out of the shops, much to Dad's disgust.

One night we camped in the mouth of a lane and the cattle were happily grazing on some dry grass with a pick of green amongst it. After we all had tea Dad, Mum, Mike and Beryl went to town for some fuel and grocery supplies and on the way back they called into a farm to spend time with friends they had made, so we had a bit of time alone in the camp. A young bloke who Les had met on the side of the road a few times came into the camp on his motorbike and stopped for a chat. He offered Les a pillion ride on the bike and they zoomed down the dirt track and came back a little later, Les all smiles. He offered us girls a ride and I happily said yes after Mary hesitated. I climbed astride the pillion seat, he gave me instructions on how to sit and off we went. No helmet in those days. My brown curly hair was flying in the breeze, my eyes watering from the wind and I loved it. We got to the crossroads where he showed me what the bike could do – okay he showed off. We zoomed back to the camp. I can still remember the excitement of driving through the moon shadows of the trees. On the way back I hoped fervently that the parents were not back as I would have got a good hiding for going off with a strange young man and rightly so I guess. Thinking about it fifty years later, I realise what a dreadful risk I took, going off into the night with a complete stranger!

When we got back to the camp, Mary saw I was in one piece so she had a ride also. Then of course I thought of Les telling the parents what we had done and I got all stressed. When the parents came back, the chap was still at the camp and we all had a cup of tea and nothing was said about the road trip. For once Dad did not mind finding a young man chatting to us when they arrived home.

The drought had moved south, so again we had trouble finding enough food for the stock and they were getting poorer. The hay in the back of the truck was primarily for the horses but the cows had the occasional bale to top them up a bit.

By now, Beryl was ten years old and was sleeping in the front of the van in my little bed, while I shared the double bed with Mary in the back of the truck. For some reason, I chose to have a makeshift bed on top of the hay. As time went by, the hay separated and I slept crookedly and eventually I hurt my back. I think this was a combination of lifting the heavy calves into the truck and the crooked bed. When I complained about my back hurting, it hurt when the evening closed in and then into the morning until it warmed up a bit, I was accused of slacking off on my workload, which was a bit mean.

Mary and Les were rounding up the stock one morning and somehow she fell off the horse and the mare kicked her in the head. Les called out to us and Dad ran to Mary who was lying in a gully a short distance away with blood running out in copious amounts. Dad called to Mum to bring the truck and they hauled her into it and rushed her to the hospital in Condobolin. They were gone for a long time and we were all worried about her, not knowing what was going on. I had to get on the horse and help Les with the stock, which left Mike and Beryl in the camp on their own. The stock were nearby, so I managed to keep an eye on the two younger siblings from a distance.

Eventually the parents came back with good news that Mary was okay but would be in hospital for a while. For the next week it was a regular trip into Condobolin to visit her during the evening visiting hours, which had to be strictly adhered to as the Matron was not amused if we turned up early. It was great to see Mary for the first time and to see she was getting better. She showed us the cut on her head that had a neat horse shoe shape and her hair had been shaved around the cut showing numerous stitches in it. We didn't count the stitches but she was a lucky girl. As her beautiful red curly hair grew back, it showed the distinct shape of a horseshoe. She learnt how to hide it with a haircut shaping around it and eventually it settled right down. This was the worst accident we had that gave us the greatest fright.

After several months the cattle were sold in Condobolin and soon after the drought broke. If Dad had held the sale off for another week or so, he would have made a lot of money but he made a financial loss instead. After this we had several small droving trips in and around Moree, working for Mr Kirby who owned *Success Park* and *Wellges*.

By now Dad was thinking seriously about owning a piggery and he made a subscription to a pig magazine. They held a monthly meeting in Warialda, about an hour's drive out of Moree, and Dad made it his business to attend these meetings when it was practical. He loved to have one of us attend with him as Mum refused to go. I went along regularly but on occasion Mary or Mike would go instead. I admit, I had more of an interest in the "young blokes" who were attending rather than the pig stories that were being bandied around.

Dad decided to settle down on his own block of land. He bought 110 acres about 10 kilometres out of Moree on the Narrabri side, where the road turned off at the abattoirs. *Strathmore* had a beautiful four bedroomed house with a fully enclosed glassed-in veranda on the northern side, a large lawn surrounding it, with lovely oleander shrubs growing on two sides. The land was broken into two paddocks and it had a small set of cattle yards near the house.

In the kitchen was an oil Aga stove, a large fridge run by electricity, and later Dad arranged for the phone to be put on and our phone number was 24D. There were about six of us sharing this line and a lot of the neighbours were curious and knew what we were doing before we knew ourselves. To ring out we had one long ring, to ring a neighbour they each had a separate call. Our number was two long rings, a short ring and another two long rings. Sometimes it was hard to distinguish the rings of the phone, so close were the ring signals.

We settled into life with a bathroom where we could have a bath or a shower that was not shared with another sibling, and a toilet with a door so no getting caught squatting behind bushes. Although life became easier in many ways I didn't ever feel as warm in my bed as I did all those years ago when I snuggled with my sisters on freezing cold nights in the open truck. Certainly life on the road was tough – but I wouldn't change a minute of it.

Epilogue

Col (Colin) married Kathleen Whiting and had five children.
Emmie and Bryon Holmes had three children.
Mary (Lorraine) married twice Harris/Pedersen and had three children.
Les never married.
Patsy (Trish) married twice Bowker/ Blackwell, no children.
Mike married twice Preston/Dee adopting three children.
Beryl does not wish for any personal information to be published.

Mary (Lorraine) died in 2002 after a five-year battle with breast cancer, aged 53.
Mum (Beryl Snr) died 2004 aged 77.
Dad (Mick/Colin Snr) died 2010 from lung cancer, aged 81.
Les died 2012 from emphysema, aged 62.

Glossary

Abo – disrespectful name for an Aborigine
Abo reserve – mission
Bark – breaking of skin, knock skin off body parts
Barcoo rot – scurvy
Bloke – man
Box – to fight (boxing match)
Boxed – mixed up, especially used for mobs of sheep
Blue – a disagreement or fight
Briar – box thorn bush
Brownie – cake containing dates and ginger
Bushies – lived their whole life in the country/bush
Camp – camp for the night or have a short rest
Chap – man or fellow
Chooks – chickens
Chiack – tease, give cheek or 'muck' around (chiacking around)
Cocky – station owners, cockatoo's or galahs
Crook – ill or dishonest
Dressed meat – butchered ready to be cut up
Dunny – toilet
Forty winks – short rest or snooze
Hot – stolen
Lollies – sweets or candy
Long paddock – public stock route
Kero – kerosene
Jackeroo – young stockman learning the trade
Out of the blue – happened suddenly
Piddly – small or weak
Pictures – movies
Pissle – penis

Pong – bad smell
Plant – drover's collective work gear
Pull your leg – tease
Reserve – common
RSL – Return Soldiers League
Sack – dismiss
Sheep break – temporary holding pen for sheep
Shied – jumped aside in fright
Show – social fair or escort stock through a station/property
Skite – boastful
Sling off – tease to embarrass
Stubble – stalks from harvested grain
Tally – count
Tanked – drunk
Turnout – collective work gear
Ute – utility
Quids – pounds £
Wagga Rug – blanket made of bags
Wallaby on the – on the track, carrying your swag
Wild – angry
Willy willy – whirlwind
Yangs – wild horses
Yanks – Americans

THE DROVER'S
Daughter

ISBN: 9781925367751	Qty	.....
RRP	AU$24.99	.....
Postage within Australia	AU$5.00	.....
TOTAL★		$....

★ All prices include GST

Name: ....

Address: ....

Phone: ....

Email: ....

Payment: ❏ Money Order ❏ Cheque ❏ MasterCard ❏ Visa

Cardholder's Name: ....

Credit Card Number: ....

Signature: ....

Expiry date: ....

Allow 7 days for delivery.

Payment to:

Marzocco Consultancy (ABN 14 067 257 390)
PO Box 12544
A'Beckett Street, Melbourne, 8006 Victoria, Australia
admin@brolgapublishing.com.au

BE PUBLISHED

Publish through a successful publisher.
Brolga Publishing is represented through:
• **National** book trade distribution, including sales,
marketing & distribution through Dennis Jones &
Associates.
• **International** book trade distribution to
> • The United Kingdom
> • North America
> • Sales representation in South East Asia
• **Worldwide e-Book distribution**

For details and inquiries, contact:
Brolga Publishing Pty Ltd
PO Box 12544
A'Beckett St VIC 8006

Phone: 0414 608 494
markzocchi@brolgapublishing.com.au
ABN: 46 063 962 443
(Email for a catalogue request)

www.ingramcontent.com/pod-product-compliance
Lightning Source LLC
Chambersburg PA
CBHW062052270326
41931CB00013B/3046